MACMILLAN
WORK OUT
SERIES

Work Out

Physics

'A' Level

The titles in this series

For examinations at 16+

- Accounting
- Biology
- Chemistry
- Computer Studies
- Economics
- English
- French
- German
- Human Biology
- Mathematics
- Numeracy
- Physics
- Sociology
- Spanish
- Statistics

For examinations at 'A' level

- Applied Mathematics
- Biology
- Chemistry
- English Literature
- Physics
- Pure Mathematics
- Statistics

For examinations at college level

- Dynamics
- Engineering Thermodynamics
- Mathematics for Economists
- Operational Research

MACMILLAN
WORK OUT
SERIES

Work Out Physics 'A' Level

T. B. Akrill
and
S. J. Osmond

First published 1986
Reprinted 1987 (with corrections); reprinted again 1987

Published by
MACMILLAN EDUCATION LTD
Houndmills, Basingstoke, Hampshire RG21 2XS
and London
Companies and representatives
throughout the world

Typeset by TecSet Ltd,
Wallington, Surrey
Printed in Hong Kong

British Library Cataloguing in Publication Data
Akrill, T. B.
Work out physics 'A' level.—(Work out series)
1. Physics—Examinations, questions, etc.
I. Title II. Osmond, S. J.
530′.076 QC32
ISBN 0-333-40820-9
ISBN 0-333-43509-5 (Export pbk)

Contents

Acknowledgements

The cover photograph, by Paul Brierley, shows a high-voltage flashover test at 33 000 volts a.c.

The author and publishers wish to thank the following who have kindly given permission for the use of copyright material:

The Associated Examining Board, Northern Ireland Schools Examinations Council, Scottish Examination Board, University of London School Examinations Board and University of Oxford Delegacy of Local Examinations for questions from past examination papers.

Every effort has been made to trace all the copyright holders but if any have been inadvertently overlooked the publishers will be pleased to make the necessary arrangement at the first opportunity.

The University of London Entrance and School Examinations Council accepts no responsibility whatsoever for the accuracy or method in the answers given in this book to actual questions set by the London Board.

Acknowledgement is made to the Southern Universities' Joint Board for School Examinations for permission to use questions taken from their past papers but the Board is in no way responsible for answers that may be provided and they are solely the responsibility of the authors.

The Associated Examining Board, the University of Oxford Delegacy of Local Examinations, the Northern Ireland Schools Examination Council and the Scottish Examination Board wish to point out that worked examples included in the text are entirely the responsibility of the author and have neither been provided nor approved by the Board.

Examination Boards for Advanced Level

Syllabuses and past examination papers can be obtained from:

The Associated Examining Board (AEB)
Stag Hill House
Guildford Surrey GU2 5XJ

University of Cambridge Local Examinations Syndicate (UCLES)
Syndicate Buildings
Hills Road
Cambridge CB1 2EU

Joint Matriculation Board (JMB)
78 Park Road
Altrincham
Cheshire WA14 5QQ

University of London School Examinations Board (L)
University of London Publications Office
52 Gordon Square, London WC1E 6EE

University of Oxford Delegacy of Local Examinations (OLE)
Ewert Place
Summertown
Oxford OX2 7BZ

Oxford and Cambridge Schools Examination Board (O & C)
10 Trumpington Street
Cambridge CB2 1QB

Scottish Examination Board (SEB)
Robert Gibson & Sons (Glasgow) Ltd
17 Fitzroy Place Glasgow G3 7SF

Southern Universities' Joint Board (SUJB)
Cotham Road
Bristol BS6 6DD

Welsh Joint Education Committee (WJEC)
245 Western Avenue
Cardiff CF5 2YX

Northern Ireland Schools Examination Council (NISEC)
Examinations Office
Beechill House
Beechill Road
Belfast BT8 4RS

Introduction

How to Use this Book

This book is not designed to replace a textbook – indeed it will be of most effect if used alongside one. It aims to give you the opportunity to practise your physics by giving examples and questions on topics that form the core of all 'A' Level syllabuses.

Each chapter begins with sections telling you:

- What you should recognise.
- What you should be able to use.
- About the most important ideas in that chapter.

These will help you prepare for the work on that chapter. If you have a textbook you should make yourself familiar with its contents and index so that you can follow up the ideas you are finding troublesome.

Your examination board has extra things in its syllabus that are not common to all the boards, so you must also at this stage look at your syllabus. Syllabuses can be obtained from the boards whose addresses are at the front of the book on pages vii-viii.

Once you have an idea of what you should have covered in any topic area you can start on the questions. There are two types:

- *Worked examples.* The answer follows immediately after the question, with additional comments.
- *Questions.* The numerical answers and hints are to be found later in the book.

The examples and questions either are from real examination papers or are specially devised to fill in gaps and avoid unnecessary repetition. They are all of 'A' Level standard, but some are of course easier than others.

The answers to the worked examples are written out a little more fully than those we would expect from a good candidate. This is to ensure clarity. The additional comments in **bold type** are written in the light of the authors' experience of both teaching and examining at 'A' Level. They would not be part of a real answer.

The examples can be used in a variety of ways, but a suggested routine is:

- Read the question carefully. So many marks are lost by careless reading that practice here is essential.
- Without looking at the answer, have a go at answering it in note form with sketch diagrams where necessary. Consult the start of the chapter and your textbook at this stage. Try the numerical problems.
- Consult the answer. Find out if and where you have gone wrong and go through it very carefully. Follow any algebra through stage by stage and beware if your answer is much longer than the model. But do not just find your mistakes – learn from them!

With each example you should gain confidence and improve your technique. Once you have studied the examples, try the questions – first without the hints, and then if necessary with their help.

The following section on revision will emphasise further the value of doing physics and not just reading textbooks and notes.

The introductions to the chapters will also help you sort out some of the problems that confuse lots of 'A' Level students:

- *Units*. The SI system is clearly laid down and used almost the world over. (See in particular page 7.) When reading the answers, look at the units carefully and use them in your own answers.
- *Abbreviations*. Physicists the world over use a consistent set of abbreviations. Even the Russians with their different alphabet use Roman and Greek letters in their formulas. A few Greek letters crop up more than most. These are α (alpha), β (beta), γ (gamma), ϵ (epsilon), λ (lambda), μ (mu – this is also used for micro, 10^{-6}, not to be confused with m for milli, 10^{-3}), θ (theta), ϕ (phi), ρ (rho) and the very familiar π (pi). Two letters δ, Δ (delta) and ω, Ω (omega) crop up in both small and capital forms. Even with the Roman letters, be sure about whether you are meant to use the small or capital form. For instance, C is used to represent a capacitance, but c represents the speed of waves.

Another couple of questions that 'A' Level students might ask are:

- What do I need to remember from GCE/CSE?

Since one piece of physics builds on another, there is not normally a great problem; you will have been using your earlier physics all through the 'A' Level course. It is, however, worth having a quick look through what you did before, just in case there are areas in which you are very rusty. Though questions are unlikely to ask specifically about them, you may find that it is easier to do an answer if you are still familiar with them. Examiners tend to assume you still remember 'O' Level or CSE.

- What if my mathematics is not strong?

If you had a good understanding of mathematics before 'A' Level you will know most of the things necessary for 'A' Level physics. But there are a few mathematical results necessary for physics that only come up in 'A' Level mathematics. The examination boards list these in their syllabuses and some physics textbooks cover them in an appendix. Time spent on these results is time well spent because the same result is used many times over in different physics topics. For instance, exponential decay formulas crop up in work on capacitors and in work on radioactive decay.

If you are having trouble with the mathematics, read through the working very carefully and ask someone to explain the steps you do not understand. If you consult a mathematics textbook remember that it may go into much more detail than you need. But do not worry too much since the boards are aware of the problem that not all candidates do both subjects at 'A' Level and try to make sure that they are testing mainly your physics and not your mathematics.

Revision

Literally, revision means 'seeing again' and it is a vital part of learning, though the word often gets used simply to mean preparation for an examination or test.

The process of learning has been studied and it has been found that the amount that can be recalled drops dramatically unless the new ideas are revised or reviewed. The graph opposite shows what happens: if an idea is reviewed after 10 minutes, then after a day and so on, it is possible to get to the stage where nearly 100% recall is achievable long after the last review. The explanation for the graph lies in the changes that take place in the cells of the brain and in the connections between them.

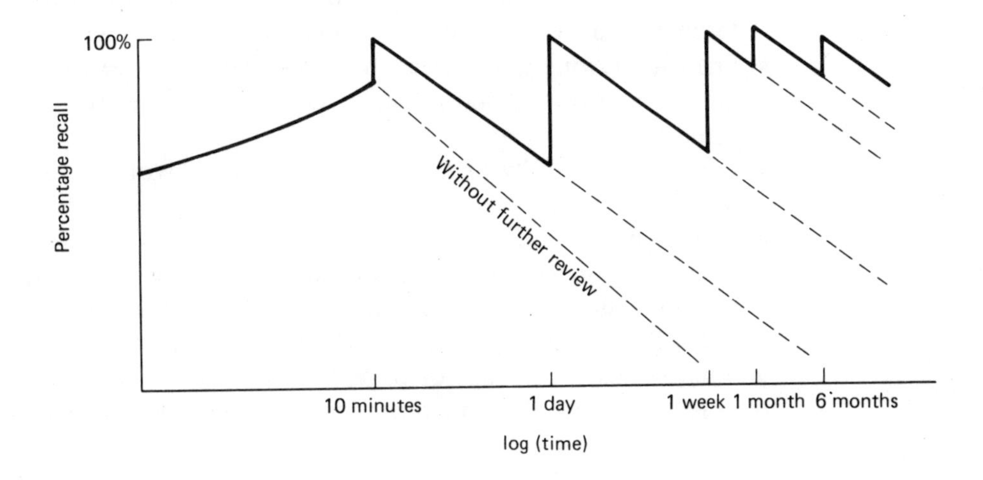

The keys to effective preparation for examinations are:

- Planning a timetable that includes time for review.
- Making revision an active process.
- Finding a suitable time and environment for study.

Let us look at these in turn.

'I haven't got time to work just now, but I'll revise all day for the last week of the holidays.'

Revision for summer examinations should aim at short but regular sessions. Little but often is the key. In a subject like physics that involves understanding concepts more than absorbing information, the early start is particularly important. A little time spent planning in advance, and at the start of each session, saves hours. Write down your plan after thinking about the following:

- What syllabus have I got to cover?
- What notes and textbooks do I have available?
- What questions have I got to practice with?
- How much time do I have available during term and during the holidays?
- Am I being over-ambitious?

You should never sit down to study without a clear idea of what you intend to do. If you finish your set tasks a little early, treat yourself to some free time.

Another hint is to plan just a very small amount of work for your first session, but make absolutely sure you get it done. You will feel you are under way and the rest of your revision will seem much less intimidating.

'It's such lovely weather that I'll take my notes outside and revise in the sunshine.'

Better to study properly for half an hour and sunbathe for two rather than pretend to revise all afternoon. Study needs to be active, and browsing through notes has little value. Things to try and do are:

- Find what the textbook has to say about what you have just read in your notes and maybe add to them or put down page references for later review.
- Do short questions to check your understanding.
- Prepare answers to longer questions in note form.
- Mark up the most important parts of your notes with underlinings or a fluorescent pen.
- Make a note of things to review at the end of the session and at the beginning of the next day's session.

- List the things that you have not quite sorted out, so that you can ask your teacher. Teachers are only too happy to help those who are helping themselves.
- Practise drawing quick clear diagrams that might be useful in answering examination questions. Pictures help recall: it is easier to remember a face than a name.
- Look very carefully at the units in numerical work and see how they are interrelated.
- Study any formula sheet that you are going to be given in the examination. You may not need to know all the formulas by heart, but you do need to be very familiar with them and know how to find them and use them.

Reading is an active process in itself. Don't try to do too much at a time. When you have chosen the page or pages you intend to cover, go through quickly and then read through the text thoroughly. Finally look through again quickly to get an overall view.

'I work much better with music on.'

This is often true when we do what are basically mechanical or boring tasks, but efficient study needs good conditions. Things to aim for are:

- A comfortable chair at a big table.
- Bright but diffuse lighting.
- No visual distractions such as a window with a view.
- Few sounds that distract.
- A temperature of about 20°C (68°F). It is better to wear a jumper than sit in too warm a room since the brain is at its sharpest at the slightly lower temperatures.
- Planned breaks, but don't let the time for every cup of coffee spin out to a half hour or so.

When you are not actually studying, get a reasonable amount of exercise and a good night's sleep.

Preparing for an important examination should not stop you doing the things you really want to do. If you plan well and avoid wasting time, you will be able to continue with most of your favourite activities. Those who talk most about the revision they need to do are generally those who are not actually doing it. Those who really are revising find it satisfying and feel they are getting somewhere as they work through their plan and find plenty of time for relaxation too.

The Examination Itself

'Didn't you say the exam started at half-past nine?'

Always make sure that you know in good time exactly where and when the examination is going to be held so that you can arrive in good time, but not too early. In a subject such as physics it is best to plan the end of your revision a day or so before the examination and relax. A good night's rest is far more valuable than a couple of extra hours of work. Literally last minute revision tends to make people panicky by reminding them of things they do not know or understand well. Have all the things you will need ready the day before so you have a chance to do something about it, if for instance your calculator is not working. A few suggestions for things you will find useful are:

- A tried and trusted pen which you have filled or for which you have refills. Many find a fountain pen more comfortable for long periods of writing.
- Spare ball points of various colours. A diagram will often get a point across more quickly and more simply if a second colour is used.
- Calculator with spare batteries or better still two calculators.

- Pencils, ruler, rubber and compasses. But remember, don't rule lines when free-hand diagrams are sufficient. Almost all the diagrams in the answers to examples in this book are in fact freehand. However bad your drawing, electrical circuit diagrams will be as clear freehand as when ruled.
- A watch or small clock.
- Don't bother with Tipp-Ex. Simple crossings out are much quicker and the marker will ignore them. He has enough to do without reading anything with a line through it.

'But you told me we only had to do four questions.'

Arrive fully prepared having eaten something. Talking about the examination with other candidates does no good at all. You only make each other nervous. Once you are given the printed question paper, read the instructions carefully. You should be familiar with the type of paper you are sitting but it is as well to be sure about how long you have and how many questions you must answer. What happens then depends on the type of paper. The common types are:

- Coded answer or multiple choice.
- Short structured questions.
- Longer questions often including descriptions and a numerical problem or analysis of data.
- Long questions of an essay answer type.
- Comprehension questions.
- Practical examinations, which will be covered in Chapter 8.

'It's only the multiple guess today.'

Though some candidates nickname multiple choice questions multiple guess, these papers require a high degree of mental effort. There is a certain amount of technique to help you get the highest mark you deserve.

- Don't rush because you have between two and three minutes per question. It is better not quite to finish than to do all the paper carelessly. Check each answer as you go, but do not leave time to check at the end. A correct answer arrived at after a minute or two's careful thought is often altered in a ten second 'check'. If a question is taking too long or you are unsure about the answer, note down its number and come back to it at the end.
- Read the questions very carefully and read each answer thoroughly even if you think you have already spotted the correct one. Be active: sketch a diagram, write down a relationship, do anything that helps you think clearly.
- In numerical questions, don't look at the answer until you have worked it out using rough paper. The wrong answers are often very plausible.
- Scribble on the question paper and rough paper as much as you like, even though you will have been told not to on old papers that have been lent to you.
- Expect questions that are trying to catch you out, sometimes by appearing too simple or by giving data you do not need.
- Do not be worried by too many of one letter since they are random.
- Be absolutely sure to get the easier questions right, by not rushing them. The questions all carry the same marks.
- Finally if in doubt guess, but remember if you have removed patently wrong answers you improve your chances. If you can whittle the choice down to two you have a 50% chance of being right.

A certain amount of practice with old papers is vital, but technique will not improve indefinitely and make up for not having learnt the physics.

There is normally no choice of questions in coded answer papers, but a choice is common in other papers and so it is vital to know how many you must do and whether they must be chosen from particular sections. Doing, say, four questions instead of five can be disastrous because you have thrown away all chance of getting 20% of the marks. Doing too many wastes time, but is not quite so serious. Keep to your time allocation for each question even if it is tempting to do otherwise.

The model answers in this book show how to answer questions, but a few hints can be summarised:

- Use simple straightforward English.
- Use clear labelled diagrams.
- Leave space so you can add things that occur to you later.
- Explain all your working and put in correct units.
- Round off your final answers to a sensible number of significant figures.
- Check your working as you go.
- Look at the marks to gauge how much detail is likely to be expected and, in the case of papers where you answer on the question paper, look at the space available. Six marks probably means six ideas are needed in the answer.

In the case of longer and more open-ended questions be sure to plan your answer first. This will avoid waffle and ensure that every sentence says something relevant and is worthy of a mark. Remember no candidate will have time to write the perfect answer, so just make sure you put down plenty of physics.

In comprehension questions read the passage quickly, look at all the questions and then read it carefully. Finally start answering the questions, referring to the passage as necessary.

If you have time at the end, check all your numerical working first, paying special attention to units and significant figures. Then go through the rest of your answers checking wording and that you really have answered the question asked. You may be able to add some extra detail.

Finally do not discuss the paper with the other candidates and do a post mortem. Some papers may suit you better than others, but whether everyone thinks the examination was hard or easy is of no interest since the pass marks are decided after the papers are marked and the difficulty of the paper is taken into account. Forget all about it and relax so as to be fresh for your next examination.

There is one piece of advice that cannot be repeated often enough to candidates in every range of ability:

- **READ THE QUESTION.**

1 Quantities and Units

1.1 You Should Recognise

(a) The Basic Quantities and Units of the SI System

Quantity	*Symbol*	*Unit*
mass	m	kilogram, kg
length	l	metre, m
time	t	second, s
current	I	ampere, A
temperature interval	T	kelvin, K
amount of substance	n	mole, mol
angle	θ	radian, rad

These units are the basic units from which all other units can be obtained. Lists of derived units, e.g. newton, N, which is a name for kg m s^{-2}, or coulomb, C, which is a name for A s, can be found on the first page of the chapter in which it is widely used: see pages 73 and 144 for example.

The only definition you are usually asked for is that of the ampere. The ampere is that constant current which, if maintained in two straight parallel conductors of infinite length, of negligible circular cross-section, and placed 1 metre apart in vacuum, would produce between these conductors a force equal to 2×10^{-7} newton per metre of length.

(b) Multiple and Submultiple Prefixes

Multiple	*Prefix*	*Symbol*
10^3	kilo	k
10^6	mega	M
10^9	giga	G
10^{12}	tera	T

Submultiple	*Prefix*	*Symbol*
10^{-3}	milli	m
10^{-6}	micro	μ
10^{-9}	nano	n
10^{-12}	pico	p

You will find these crop up regularly with certain units: e.g. μF, microfarad or 10^{-6} F, is very common when describing a capacitor; nm, nanometre or 10^{-9} m, when giving an optical wavelength; kPa, kilopascal or 10^3 Pa (10^3 N m^{-2}), when quoting a gas pressure.

1.2 You Should be Able to Use

(a) Equations to Define Physical Quantities

Any quantity, other than those listed in section 1.1(a) above, can be defined by using a word equation, e.g.

$$\text{density} = \frac{\text{mass}}{\text{volume}}, \qquad \rho = \frac{m}{V}$$

$$\text{momentum} = (\text{mass})\,(\text{velocity}), \qquad p = mv$$

$$\text{resistivity} = \frac{(\text{resistance})\,(\text{cross-sectional area})}{\text{length}}, \qquad \rho = \frac{RA}{l}$$

These defining equations, which are given at the beginning of the chapters in which they are used frequently, also fix the unit for the derived quantity. You should appreciate that the unit for density must be kg m^{-3}, for momentum kg m s^{-1} (which is equivalent to N s) and for resistivity Ω m^2 m^{-1} or simply Ω m.

(b) Units or Dimensions to Test Possible Relationships

As any equation in physics must have the same units on each side, it is possible to detect an impossible relationship by considering only its units. For example, suppose you seem to remember that the drag force, F, on a tennis ball moving at a speed v through air of density ρ is given by $F = \frac{1}{2}\rho v$.

As ρ has the unit kg m^{-3} and v the unit m s^{-1}, the right-hand side has units

$$\frac{\text{kg}}{\text{m}^3} \times \frac{\text{m}}{\text{s}} = \frac{\text{kg}}{\text{m}^2\,\text{s}} = \text{kg m}^{-2}\ \text{s}^{-1}$$

while the left-hand side has units

$$\text{N} = \text{kg m s}^{-2}$$

so the equation you have tried to remember *must be wrong*. ($F = \frac{1}{2}\rho v^2$ has the correct units; but note that the $\frac{1}{2}$ could be wrong – we can never be sure a formula is right by this method.) If you use dimensional symbols M, L and T rather than units kg, m and s to check equations in this way, be careful not to muddle M for mass with m for metre, i.e. length.

The contents of ( ) in ln( ), $\log_{10}$( ) and exp( ) must be dimensionless, that is they are simply numbers. Trigonometric functions such as sine or cosine are also numbers.

(c) Conventions for Labelling Graph Axes and Heading Tables of Data

The important thing to remember here is that a quantity in physics is always represented by a number *and* its unit. A pressure might be

$p = 6.4 \times 10^5$ N m^{-2} or 640 kN m^{-2} or 640 kPa

A graph of p against something would have (i) $p/10^5$ N m^{-2} or (ii) p/kPa on its axis and the number plotted would be (i) 6.4 or (ii) 640. Similarly if a time period $T = 0.74$ s is measured then $T^2 = 0.55$ s^2 and you would have graph axes of (i) T/s or (ii) T^2/s^2 and plot the numbers (i) 0.74 or 0.55.

Two graphs are given in Fig. 1.1 and some tables are given below as examples; others can be found elsewhere in the book.

Density / $\frac{\text{kg}}{\text{m}^3}$	$\rho/10^3$ kg m^{-3}
7870	7.87
8930	8.93
2710	2.71

Current/A	$I/10^{-2}$ A	ln(I/mA)
0.042	4.2	3.74
0.61	6.1	4.11
0.079	7.9	4.37

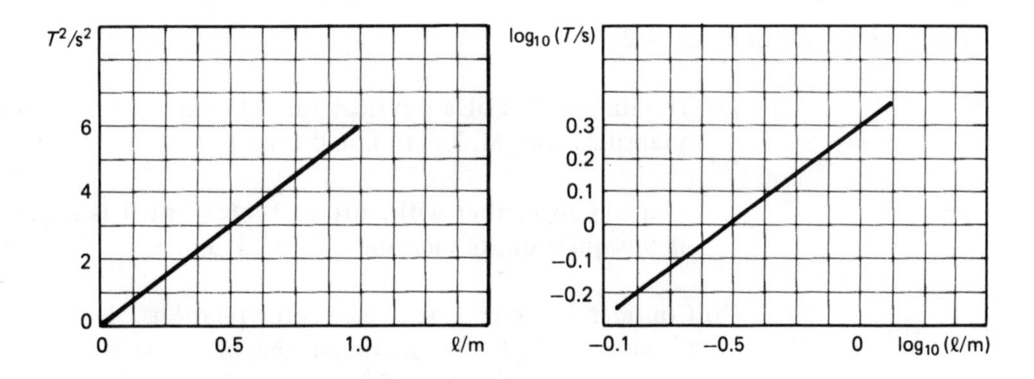

Figure 1.1

1.3 Worked Examples

Example 1.1

If the Earth is taken to be a uniform sphere of radius r and density ρ, the gravitational intensity at the surface is given by

$g = kr\rho$,

where k is a constant. What are the units of k?

A kg m s^{-2} **B** kg m^{-1} s^{-1} **C** kg m^{-3} s^2 **D** kg^{-1} m^3 s^{-2}
E kg^{-1} m^2 s^{-2} (OLE)

Solution 1.1

g has units $\dfrac{m}{s^2}$

$r\rho$ has units $(m)\left(\dfrac{kg}{m^3}\right) = \dfrac{kg}{m^2}$

$\therefore$ k must have units $\left(\dfrac{m}{s^2}\right) \div \left(\dfrac{kg}{m^2}\right) = \left(\dfrac{m}{s^2}\right)\left(\dfrac{m^2}{kg}\right)$

i.e. kg^{-1} m^3 s^{-2}

Answer D

It is better to write the units as fractions rather than use the index notation in manipulating units algebraically as here, i.e. use $\dfrac{m}{s^2}$ not m s^{-2}.

Example 1.2

(a) What are meant by the *dimensions* of a physical quantity? [2]
(b) With the aid of a suitable example, explain how dimensions can be used to check the validity of an equation connecting physical quantities. [3]
(c) Given that the speed c of longitudinal waves in a solid depends only on the Young modulus E of the material and its density ρ, use the method of dimensions to derive a relationship between c, ρ and E. [5]
(OLE)

Solution 1.2

(a) The dimensions of a physical quantity express how it is related to the basic quantities mass M, length L and time T.

You can go further with current I, etc., but it is quite usual, as in the rest of this question, to mention only M, L and T.

(b) Consider $E = mc^2$, the mass–energy relation. Dimensions of E are ML^2T^{-2} (from $E = Fs$, $F = ma$). Dimensions of m are M, and of c^2 are L^2T^{-2}, so mc^2 has dimensions of ML^2T^{-2} and the equation checks.

If you are not used to using dimensions then use units kg, m and s.

(c) Suppose $c = kE^x\rho^y$ where k is a number.
The dimensions of c are LT^{-1}
of E are $ML^{-1}T^{-2}$ ($Pa = N\ m^{-2}$)
of ρ are ML^{-3}

Therefore

$$LT^{-1} = (ML^{-1}T^{-2})^x(ML^{-3})^y$$
$$= M^xL^{-x}T^{-2x}M^yL^{-3y}$$

i.e.

$$LT^{-1} = M^{x+y}L^{-x-3y}T^{-2x}$$

from which

$$x + y = 0$$
$$-x - 3y = 1$$
$$-2x = -1$$
$$\Rightarrow\ x = \tfrac{1}{2},\ y = -\tfrac{1}{2}$$

and so a possible relationship is

$$c = k\sqrt{\frac{E}{\rho}}$$

In fact $k = 1$. Again units could be used instead of dimensions M, L and T. Familiarity with a range of formulas can help you spot possible mistakes when a strange-looking answer crops up.

Example 1.3

Show that the equation relating the current density, J, in a wire to the drift speed, v, of the electrons

$$J = nve$$

where e is the charge of an electron and n is the number density of the electrons, is dimensionally correct.

Solution 1.3

You could use dimensions M, L, T and I here or, more simply, units kg, m, s and A. We shall use units.

Units of J are $\frac{A}{m^2}$ (current per unit cross-section)

of n are $\frac{1}{m^3}$ (number of electrons per cubic metre) of v are $\frac{\text{m}}{\text{s}}$

of e are C

$\therefore$ nve has units $\left(\frac{1}{\text{m}^3}\right)\left(\frac{\text{m}}{\text{s}}\right)(\text{C}) = \frac{\text{C}}{\text{s m}^2} = \frac{\text{A}}{\text{m}^2}$

which are the units of J.

Example 1.4

The Planck constant relates the energy associated with a quantum of radiation to the frequency. It has dimensions

A MLT^{-1} **B** MLT^{-2} **C** ML^2T^{-1} **D** ML^2T^{-2}
E ML^2T^{-3} (NISEC)

Solution 1.4

The Planck constant is defined in $E = hf$, so h has the dimensions of E/f.
Dimensions of energy are ML^2T^{-2} ($E = Fs, F = ma$)
of frequency are T^{-1}

$\therefore$ E/f has dimensions $\text{ML}^2\text{T}^{-2} \div \text{T}^{-1}$
i.e. ML^2T^{-1}

Answer **C**

Or you might remember or see in your formula sheet that the unit of h is J s.

Example 1.5

Show that the units of the product RC (resistance × capacitance) reduce to seconds. [4]

Solution 1.5

The unit of R is the ohm, $\Omega = \frac{\text{V}}{\text{A}}$ ($V = IR$)

The unit of C is the farad, $\text{F} = \frac{\text{C}}{\text{V}}$ ($Q = VC$)

$\therefore$ The units of RC are $\Omega\,\text{F} = \frac{\text{V}}{\text{A}} \times \frac{\text{C}}{\text{V}} = \frac{\text{C}}{\text{A}}$

But coulomb C = A s

$\therefore$ The units of RC are $\frac{\text{C}}{\text{A}} = \frac{\text{A s}}{\text{A}} = \text{s}$

1.4 Questions

Question 1.1

Determine the units of the following physical quantities in terms *only* of the four basic SI units kg, m, s and A.

(a) power,

(b) resistance, and

(c) the universal gravitational constant G.

Start with a definition or physical law in each case, and show your reasoning. [7]

Question 1.2

An electric fan, with effective cross-sectional area A, accelerates air of density ρ to a speed v. What is the power needed for this process?

A $\rho A v$ **B** $\frac{1}{2}\rho A v$ **C** $\rho A v^2$ **D** $\frac{1}{2}\rho A v^2$ **E** $\rho A v^3$

(OLE)

1.5 Answers to Questions

1.1 (a) kg m^2 s^{-3}, (b) kg m^2 s^{-3} A^{-2}, (c) kg^{-1} m^3 s^{-2}.
To get from (a) to (b) is easy, but you should try (b) from the definition of an ohm as a volt per ampere.

1.2 This can almost be solved using units only.
Find the units of $\rho A v$ and the unit of power. This will eliminate three of the options. The final choice depends on whether a full proof depends on k.e. with its $\frac{1}{2}$, or not.
Answer *E*

2 Mechanics and Gravitation

2.1 You Should Recognise

Quantity	*Symbol*	*Unit*	*Comment*
distance along a path	s	metre, m	a scalar
displacement	s	metre, m	a vector
speed, velocity	v	metre per second, m s^{-1}	
acceleration	a	metre per second squared, m s^{-2}	
free fall acceleration	g	metre per second squared, m s^{-2}	at the Earth's surface, $g \approx 10$ m s^{-2}
momentum	p	kilogram metre per second, kg m s^{-1}	strictly, ***linear*** momentum, 1 kg m s^{-1} = 1 N s
force	F	newton, N	1 N = 1 kg m s^{-2}
weight	W	newton, N	
resultant force	F_{res}	newton, N	
energy	E	joule, J	1 J = 1 N m
gravitational potential energy	E_p	joule, J	
kinetic energy	E_k	joule, J	
power	P	watt, W	1 W = 1 J s^{-1}
moment of force	M	newton metre, N m	
torque, couple	T	newton metre, N m	
angular displacement	θ	radian, rad	dimensionless quantity
angular velocity	ω	radian per second, rad s^{-1}	1 rev s^{-1} = 2π rad s^{-1}
gravitational constant	G	newton metre squared per kilogram squared, N m^2 kg^{-2}	a constant = 6.67×10^{-11} N m^2 kg^{-2}
gravitational field strength	g	newton per kilogram, N kg^{-1}	at Earth's surface, $g_0 = 9.8$ N kg^{-1}
gravitational potential	V	joule per kilogram, J kg^{-1}	

Not listed are quantities, such as mass, length and time, which occur in all chapters and which are given in the list on page 7.

2.2 You Should be Able to Use

- Addition and resolution rules for vectors such as velocity and force.
- Expressions for velocity:

average velocity $v = \dfrac{s}{t}$ instantaneous velocity $v = \dfrac{ds}{dt}$

- Expressions for uniformly accelerated motion:

displacement $s = \left(\frac{u+v}{2}\right) t$ acceleration $a = \frac{v-u}{t}$

which also yield $s = ut + \frac{1}{2}at^2$ and $v^2 = u^2 + 2as$
for free fall from rest $s = \frac{1}{2}gt^2$

- Dynamic and static systems of forces:

Newton's second law of motion $\frac{d(mv)}{dt} = F_{res}$

or, for bodies of constant mass, $ma = F_{res}$

the principle of moments for bodies in equilibrium
Σ (clockwise moments) = Σ (anticlockwise moments) about any point

- Linear momentum:
conservation of momentum Σmv (before) = Σmv (after) any collision or explosion

the impulse–momentum equation $Ft = \Delta(mv)$

- Energy and its conservation:
work done by a force or energy transfer = (average force) (displacement)
$W = Fs$
kinetic energy k.e. = $\frac{1}{2}mv^2$
gravitational potential energy g.p.e. = mgh (close to Earth's surface)

power as rate of energy transfer $P_{av} = \frac{W}{t}$

or as (force) (velocity) $P = Fv$

efficiency of a mechanical system $\eta = \frac{\text{power out}}{\text{power in}}$

the work–energy equation $Fs = \Delta(\frac{1}{2}mv^2)$

- Uniform circular motion:

angular velocity $\omega = \frac{\theta}{t} = \frac{2\pi}{T}$

speed and angular velocity $v = r\omega$

centripetal acceleration = $\frac{v^2}{r} = r\omega^2$

- Gravitation:

Newton's inverse square law of force $F = G\,\frac{m_1 m_2}{r^2}$

gravitational field strength $g = \frac{F}{m}$ for the Earth $g = G\,\frac{m_E}{r^2}$

for bodies in free fall $\quad \Delta(\text{k.e.}) = -\Delta(\text{g.p.e.})$

gravitational potential for the Earth $\quad V = -\dfrac{Gm_E}{r}$

2.3 Describing Motion

(a) Vectors

Velocities are added according to the parallelogram law. A velocity can be resolved in two directions – see Fig. 2.1. Forces can be dealt with in the same way.

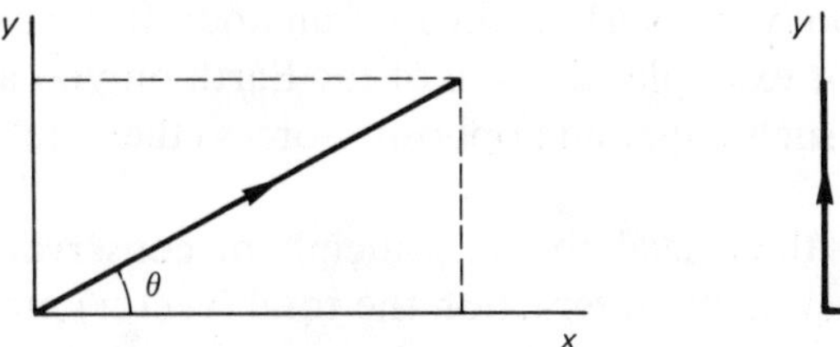

Figure 2.1

In equations linking vectors, such as $ma = F_{res}$, the direction of the two vector quantities, here a and F_{res}, is the same.

(b) Motion Graphs

A set of motion graphs for an object thrown vertically upwards is shown in Fig. 2.2. Its initial upward velocity at time $t = 0$ is v_0 and its downward acceleration is g.

The slope of the displacement–time graph is equal to the velocity; the slope of the velocity–time graph is equal to the acceleration. The area between a velocity–time graph and the time axis is equal to the displacement. All motion graphs are linked in these ways.

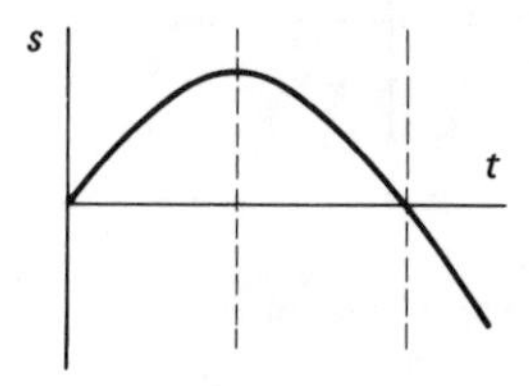

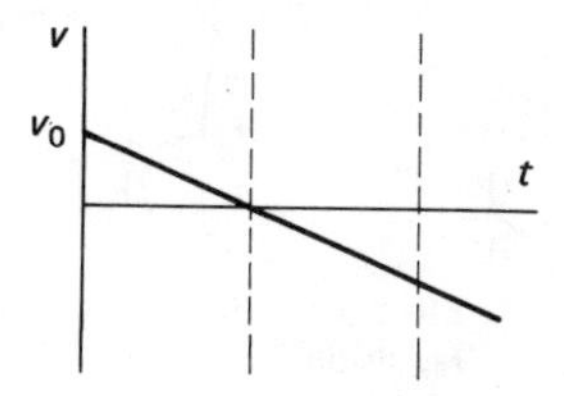

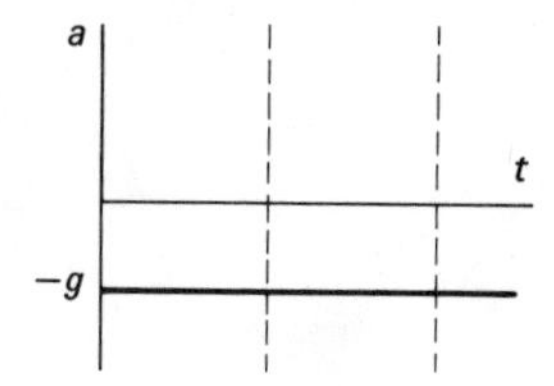

Figure 2.2

(c) Uniform Circular Motion

This involves motion at a constant speed but with a continuously varying velocity and hence it is accelerated motion. The acceleration is towards the centre of the circle and of constant size v^2/r.

2.4 Forces

(a) Dynamics

Newton's Laws of Motion

First law: a force is required to change the velocity of a body. A body moving with a constant velocity, perhaps a body that is at rest, $v = 0$, therefore has zero resultant force acting on it. It is said to be in equilibrium.

Second law: the rate of change of momentum of a body is equal to the resultant force acting on it. When using this law always draw a force diagram (a free-body diagram) of the object that is accelerating.

Third law: if body A exerts a force $+F$ on body B, then body B exerts a force $-F$ on body A. For example, the pull of the Earth on you and the pull of you on the Earth are two such equal and opposite forces (they are both gravitational forces).

Laws two and three lead to the principle of conservation of linear momentum, which states that in any interaction the total (vector) momentum is constant.

(b) Forces in Equilibrium

For a body in equilibrium:

(i) the (vector) sum of the forces, in any direction, is zero;

(ii) the sum of the moments, about any axis, is zero.

The first condition can be tested by drawing a force polygon. The polygon, a triangle in Fig. 2.3, must close when the body is in equilibrium.

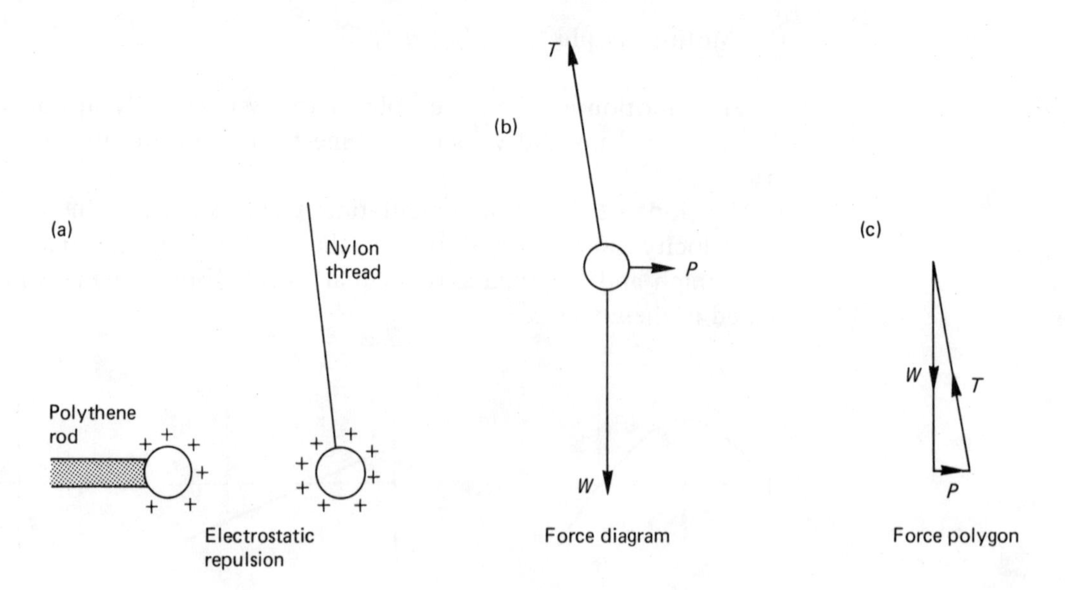

Figure 2.3

2.5 Mechanical Energy

A body possesses energy if it can do work. The mechanical energy it transfers is calculated as average force times displacement, the displacement being measured in the direction of the force. This sort of energy transfer is also called work.

Mechanical energy can be (i) kinetic energy, k.e., (ii) gravitational potential energy, g.p.e. or (iii) elastic potential energy, e.p.e., which is sometimes called strain energy. Sound energy is also mechanical energy. Mechanical energy is sometimes conserved in systems where friction effects are negligible. In general total energy is conserved, the mechanical energy being transferred from chemical or electrical energy, etc., and being transferred to internal or electrical energy, etc.

2.6 Gravitation

All bodies exert gravitational forces on one another. A large mass, like the Earth, produces a gravitational field. Near its surface this g-field is nearly uniform but on a larger scale it is radial (see Fig. 2.4) and is an inverse-square-law field, $g \propto 1/r^2$, where r is measured from the centre of the Earth. Gravitational fields are described using lines of force. Perpendicular to these lines are equipotential surfaces.

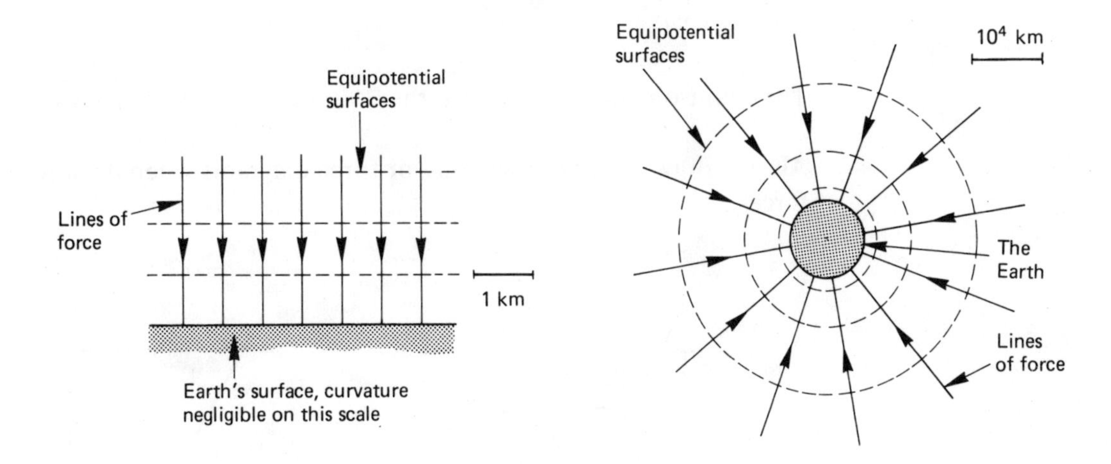

Figure 2.4

Differences of g.p.e. in the uniform-field case are given by $mg\Delta h$; the difference in g.p.e. per unit mass is called the gravitational potential difference, ΔV.

Satellites move in the Earth's field where the sum of their k.e. and g.p.e. is constant. If they move in a circular orbit, then

$$m\frac{v^2}{r} = G\frac{mm_E}{r^2} \qquad v = \sqrt{\frac{Gm_E}{r}}$$

i.e. there is a unique v for a given r.

2.7 Worked Examples

Example 2.1

A 2 kg mass is hanging by a string from the roof. A horizontal force H is applied to hold the string in the position shown in the diagram.

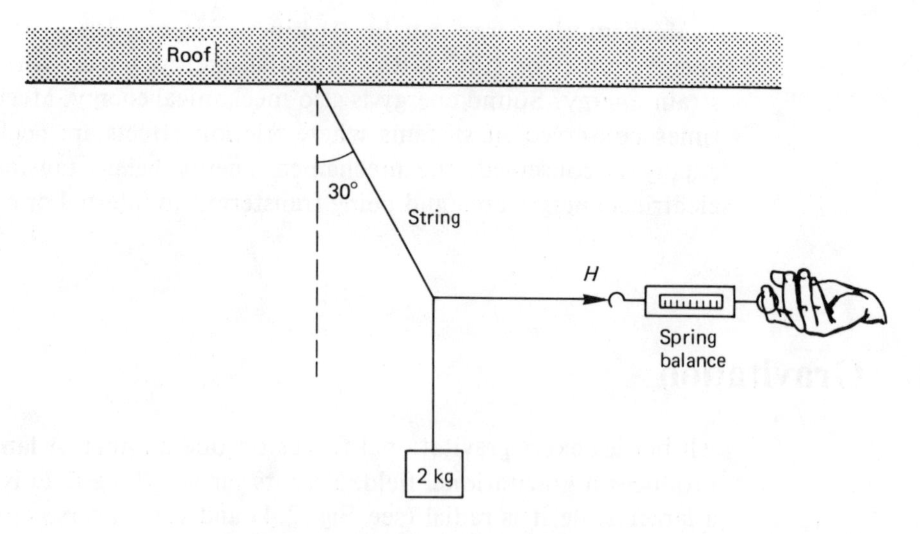

By drawing a scale diagram or otherwise, find the reading on the spring balance. [3]

(SEB)

Solution 2.1

The point on the string where the 2 kg mass is pulled sideways is in equilibrium.

Always draw separate free-body or force diagrams when dealing with problems on forces.

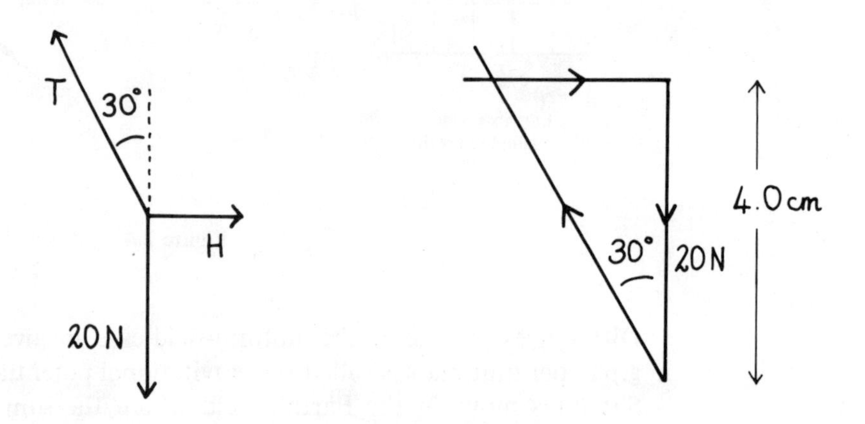

Solution 1 – scale drawing:

Scale 10 N ≡ 2 cm

H measures 2.4 cm

$\therefore\ H = 12$ N

The force polygon, a triangle here, always closes when the body is in equilibrium.

Solution 2 – calculation:

Resolving horizontally $H = T \sin 30°$

Resolving vertically $20\ \text{N} = T \cos 30°$

$$\Rightarrow \quad \frac{H}{20\ \text{N}} = \frac{\sin 30°}{\cos 30°} = \tan 30°$$

so that $H = (0.58)\,(20\ \text{N}) = 12$ N (2 sig. fig.)

Example 2.2

The diagrams show a dinghy that is sailing into the wind. The wind blowing on the sail produces a force tangential to the sail (T) and a force at right angles to the sail (P).

(a) Explain how these two forces can produce a forward force along the fore and aft line of the dinghy. [3]

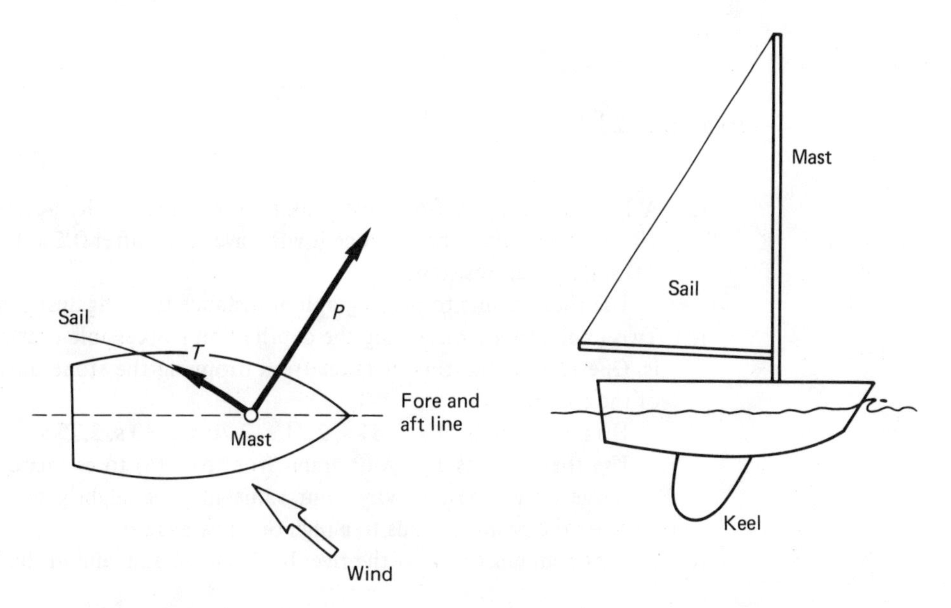

(b) After thinking about the forces on the sail give *two* reasons for having a heavy keel. Explain them carefully. [3]

Solution 2.2

(a) Resolving the forces P and T on the sail into two components:
(i) across the fore and aft line
(ii) along the fore and aft line gives

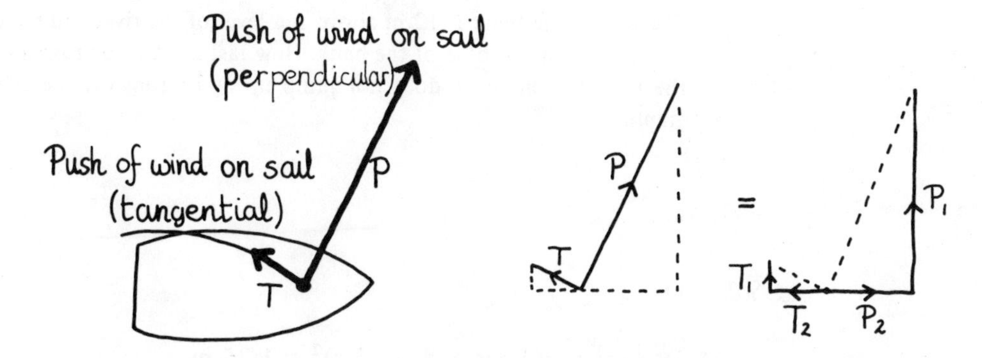

So long as $P_2 > T_2$ there will be a resultant force forwards along the fore and aft line.

This is all about forces as vectors which can be resolved using the parallelogram rule. Diagrams make the best explanations.

(b) The resultant sideways force at right angles to the fore and aft line is $T_1 + P_1$; it will act well up the mast. The keel
(i) produces a force in the opposite sense to $T_1 + P_1$, which prevents the dinghy slipping sideways (making leeway), and

(ii) when the dinghy tilts, the keel's weight produces a moment which balances the moment produced by $T_1 + P_1$ and thus prevents the dinghy toppling over.

Example 2.3

(a) A body is dropped from rest. Taking the value of the acceleration due to gravity as 10 m s^{-2} calculate the distance it will have fallen after 0.5 s, 1.0 s, 1.5 s, 2.0 s, 2.5 s and 3.0 s. Ignore air resistance.

Use these results to plot a graph of distance fallen against time. [6]

(b) Two explorers are measuring the depth of an inaccessible chasm by dropping stones into it. One records the time it takes from dropping the stone until he hears it hit the floor of the chasm.

His results are 3.71 s, 2.81 s, 2.71 s, 2.70 s, 2.80 s, 2.75 s.

Use these results and your graph from part (a) to estimate the depth of the cavern.

Suggest two reasons why your value may be slightly wrong and say in each case whether the problem leads to a high or a low answer. [6]

(c) A stunt man must run off the river bank shown and land in the water.

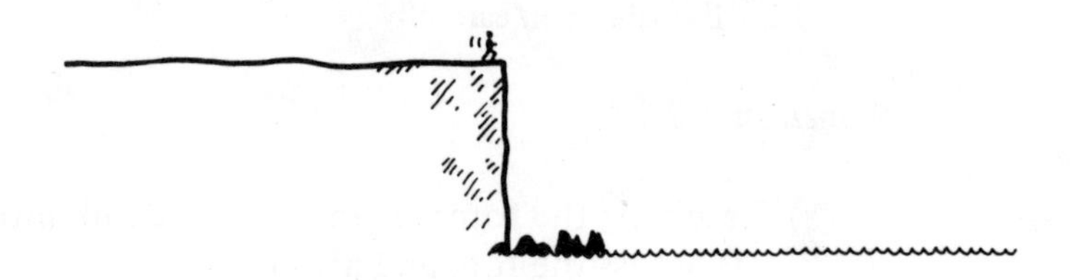

The top of the bank is 12 m above the level of the river and rocks extend 6 m out into the river from the base of the bank. How fast must he be running if he is not to land on the rocks? Assume he does not jump up as he runs off the edge. You may use your graph. [4]

Solution 2.3

(a) $s = \frac{1}{2}gt^2 = \frac{1}{2}(10 \text{ m s}^{-2})(0.5 \text{ s})^2 = 1.25 \text{ m}$

Time from release	1.0 s	1.5 s	2.0 s	2.5 s	3.0 s
Distance fallen	5.0 m	11.25 m	20.0 m	31.25 m	45.0 m

The other figures are worked out in the same way or by multiplying the first answer by 4, 9, 16, 25 and 36.

(b) The first reading does not agree with the others and so is missed out in taking an average. The average time is 2.75 s and from the graph this gives a depth of 38 m.

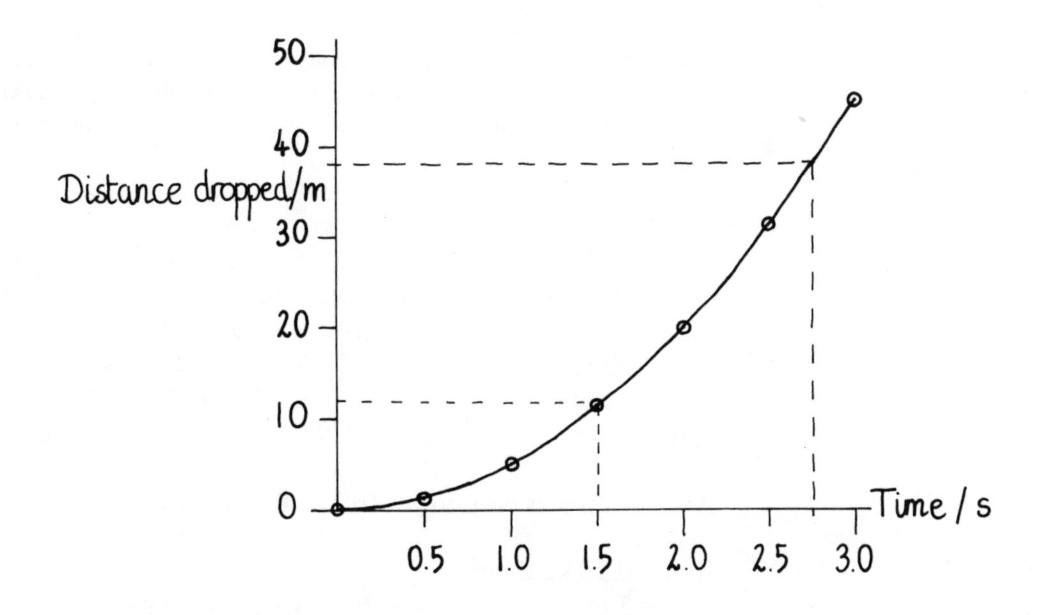

Make it obvious how you get your results from graphs.

Air resistance will cause the stone to travel more slowly and so the chasm will be less deep than 38 m. That is, the answer is too high.

The sound will take about 0.1 s to reach the top, giving a time that is too long and an answer that is too high.

(c) The graph shows that a 12 m fall takes 1.5 s. In that time the stunt-man must move 6 m in the horizontal direction.

$$\text{Minimum horizontal velocity} = \frac{6\ \text{m}}{1.5\ \text{s}} = 4\ \text{m s}^{-1}$$

This is about 10 m.p.h. It is useful to know roughly what m s^{-1} are in more familiar units.

Example 2.4

A force F acts on a body which is initially at rest. The force is always in the same direction but varies in size as shown in the graph.

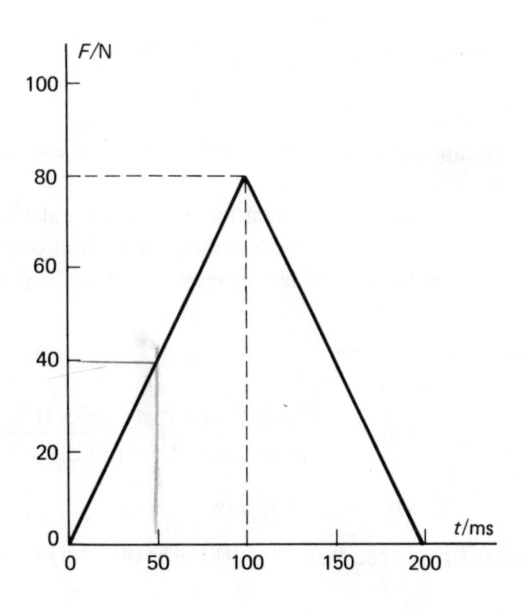

(a) Calculate the momentum of the body after 50 ms, 100 ms, 150 ms and 200 ms.
(b) If the body has a constant mass of 200 g sketch a graph showing how its velocity changes during the 200 ms in which the force acts and comment on the shape of the graph. [10]

Solution 2.4

(a) Change of momentum = (average force) (time)

after 50 ms $mv_1 = \frac{1}{2}(40\text{ N})(50 \times 10^{-3}\text{ s}) = 1.0\text{ N s}$
after 100 ms $mv_2 = \frac{1}{2}(80\text{ N})(100 \times 10^{-3}\text{ s}) = 4.0\text{ N s}$
after 150 ms $mv_3 = 4.0\text{ N s} + 4.0\text{ N s} - 1.0\text{ N s} = 7.0\text{ N s}$
after 200 ms $mv_4 = 7.0\text{ N s} + 1.0\text{ N s} = 8.0\text{ N s}$

Average force is easy to get from a linear graph.

(b) If $m = 0.20$ kg then
$v_1 = 5\text{ m s}^{-1}$, $v_2 = 20\text{ m s}^{-1}$, $v_3 = 35\text{ m s}^{-1}$ and $v_4 = 40\text{ m s}$

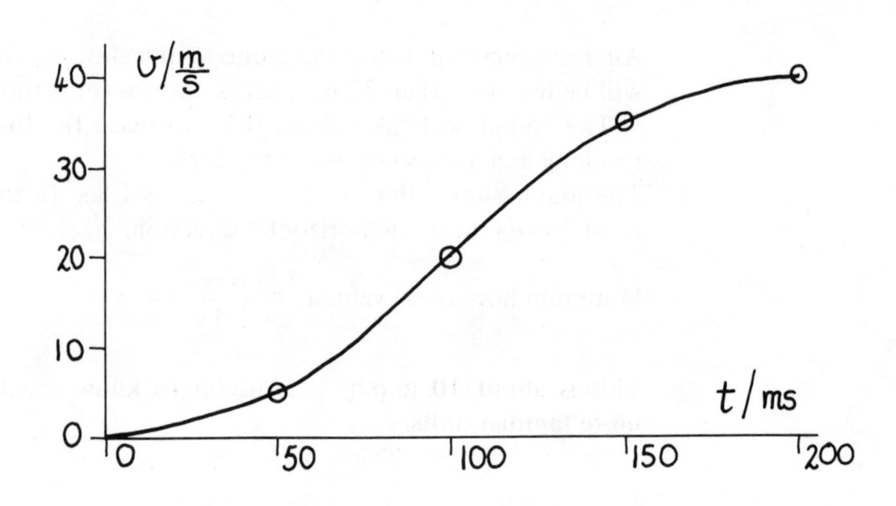

The two halves of the graph 0–100 ms and 100–200 ms are each curves, as v depends on the area under the F–t graph.

Example 2.5

A supertanker of mass 400 000 tonnes takes one hour to slow down and stop from a cruising speed of 4.0 m s^{-1}.
Assuming the acceleration is constant, calculate the acceleration of the tanker.
Hence calculate the retarding force produced by the drag and the engines operating astern.
How far will the tanker move while it is slowing down? [7]

Solution 2.5

$$\text{Acceleration} = \frac{-\text{ initial coasting velocity}}{\text{time to slow down and stop}} = \frac{-4.0\text{ m s}^{-1}}{3\,600\text{ s}} = -1.1 \times 10^{-3}\text{ m s}^{-2}$$

Using Newton's second law:

retarding force = ma = (400 000 000 kg) $(1.1 \times 10^{-3}\text{ m s}^{-2}) = 4.4 \times 10^5$ N

Because the acceleration is constant the average speed while the tanker is slowing down is $\frac{1}{2} \times 4 \text{ m s}^{-1} = 2 \text{ m s}^{-1}$.

Distance travelled = $(2 \text{ m s}^{-1})(3600 \text{ s}) = 7200$ m or 7.2 km

Example 2.6

An investigation is carried out to test the frictional force provided by car tyres.

A car of mass 600 kg is parked on a ramp and its wheels are locked so that they cannot turn. One end of the ramp is raised until the car is on the point of slipping. This occurs when the ramp is at an angle of 42°.

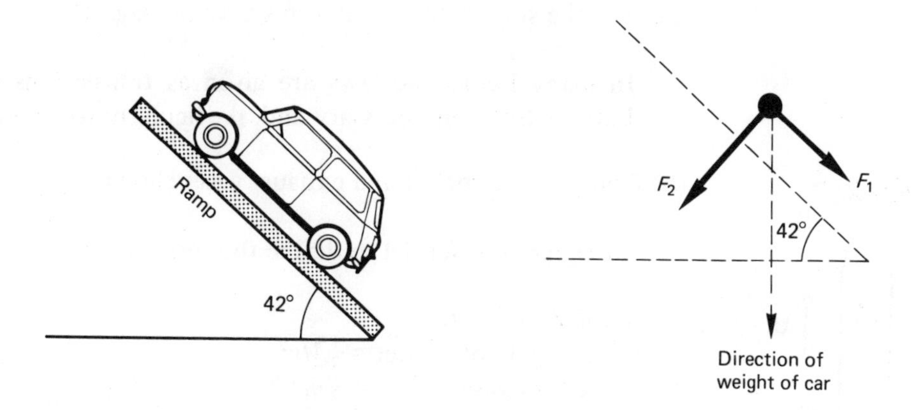

The component of the weight of the car acting parallel to the ramp is F_1.
The component of the weight of the car acting at right angles to the ramp is F_2.

(a) Calculate the value of F_1.
(b) Calculate the value of F_2.
(c) What is the direction and magnitude of the frictional force acting on the tyres?

[4]
(SEB)

Solution 2.6

(a) Weight of car = $(600 \text{ kg})(10 \text{ N kg}^{-1}) = 6000 \text{ N} = 6.0 \text{ kN}$
$F_1 = 6.0 \text{ kN} \times \sin 42° = 4.0 \text{ kN}$
(b) $F_2 = 6.0 \text{ kN} \times \cos 42° = 4.5 \text{ kN}$
(c) The car is not accelerating parallel to the ramp so it is in equilibrium. Therefore the component of its weight parallel to the ramp is equal to the frictional push of the ramp on the tyres. The force is 4.0 kN.

Example 2.7

(a) State Newton's law of motion. Explain how the *newton* is defined from these laws.
[5]
(b) A rocket is propelled by the emission of hot gases. It may be stated that both the rocket and the emitted hot gases each gain kinetic energy and momentum during the firing of the rocket.

Discuss the significance of this statement in relation to the laws of conservation of energy and momentum, explaining the essential difference between these two quantities. [5]

(c) A bird of mass 0.5 kg hovers by beating its wings of effective area 0.3 m^2.

(i) What is the upward force of the air on the bird?

(ii) What is the downward force of the bird on the air as it beats its wings?

(iii) Estimate the velocity imparted to the air, which has a density of 1.3 kg m^{-3}, by the beating of the wings.

Which of Newton's laws is applied in each of (i), (ii) and (iii) above? [8]

(L)

Solution 2.7

(a) See the statement of Newton's laws on page 16.

In many books the laws are given as translations from Newton's original Latin – this is unnecessary and, particularly for the third law, confusing.

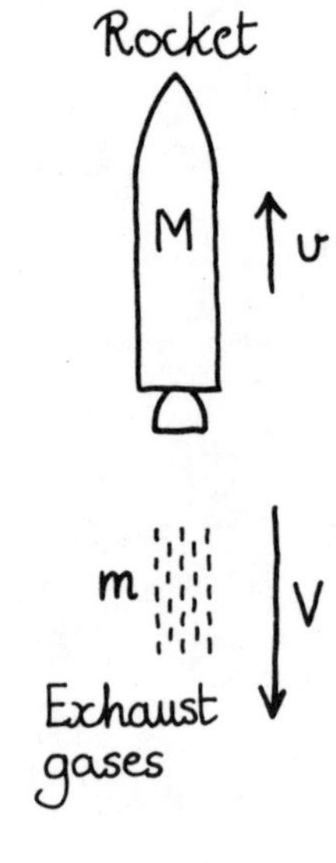

(b) Consider the rocket and exhaust gases shown.

A diagram is essential to define the terms here.

(i) Kinetic energy:

k.e. gain of rocket = $\frac{1}{2}Mv^2$

k.e. of gases = $\frac{1}{2}mV^2$

net gain k.e. = $\frac{1}{2}Mv^2 + \frac{1}{2}mV^2$

All of this comes from the chemical energy released in the explosion of the gases in the rocket motor. Energy, a scalar, is conserved.

(ii) Momentum:

momentum gain of rocket = Mv up

momentum of gases = mV down

net gain of momentum = $mV - Mv = 0$

Momentum, a vector, is conserved.

(c) (i) Pull of Earth on bird = 5 N down

So push of air on bird = 5 N up

to preserve equilibrium, Newton's first law.

(ii) Push of bird on air = 5 N down,

Newton's third law.

(iii) The downward rate of change of momentum of the air = 5 N,

Newton's second law.

Suppose its velocity is v, then in time t a volume of air = (0.3 m^2) vt is set in motion downwards:

mass of this air = (1.3 kg m^{-3}) (0.3 m^2) vt

momentum of this air = (1.3 kg m^{-3}) (0.3 m^2) vtv

rate of change of momentum of this air = (1.3 kg m^{-3}) (0.3 m^2)v^2 = 5 N

$$\text{i.e. } v^2 = \frac{5\text{ N}}{0.39\text{ N m}^{-1}} \quad \Rightarrow \quad v = 3.6\text{ m s}^{-1}$$

As this is an estimate you might give the final answer only to 1 sig. fig. You are not asked what assumptions are made – the key assumption is that a column of air of cross-section 0.3 m^2 is projected downwards all at v. It is best to give the laws as and when you use them; i.e. read the whole question before starting an answer.

Example 2.8

(a) What is the principle of conservation of momentum? [2]
(b) Explain whether the principle applies in the cases of elastic collisions and inelastic collisions and what is meant by the terms elastic and inelastic. [4]
(c) The jet of water from a hose has a cross-sectional area of 600 mm^2 and a speed of 15 m s^{-1} when it strikes a wall at right angles and is brought to rest. What is the force exerted on the wall by the jet of water?
(The density of water is 1000 kg m^{-3}.) [4]

Solution 2.8

(a) In an isolated system the total momentum after a collision is the same as the total momentum before.

Isolated means that there are no forces on bodies in the system caused by bodies outside the system. A system can be isolated for just, say, the horizontal forces and momentums.

(b) An elastic collision is one where the total kinetic energy is the same afterwards as before. An energy transformation takes place in inelastic collisions, kinetic energy being changed into other types of energy. Momentum is conserved in all isolated systems regardless of energy changes. It is a direct consequence of Newton's laws of motion.
(c) A jet of water 15 m long hits the wall each second.
Rate at which water hits the wall
$= (1000 \text{ kg m}^{-3})(15 \text{ m s}^{-1})(600 \times 10^{-6} \text{ m}^2) = 9.0 \text{ kg s}^{-1}$
Rate of change of momentum of the water at wall
$= (9.0 \text{ kg s}^{-1})(15 \text{ m s}^{-1}) = 1.4 \times 10^2 \text{ kg m s}^{-2}$
By Newton's laws the push of the water on the wall is 140 N.

Example 2.9

A large cardboard box of mass 0.75 kg is pushed across a floor by a horizontal force of 4.5 N. The motion of the box is opposed by (i) a frictional force of 1.5 N between the box and the floor, and (ii) an air resistance force kv^2, where $k = 6.0 \times 10^{-2}$ kg m^{-1} and v is the speed of the box in m s^{-1}.

Draw a free-body diagram to represent all the forces which act on the moving box. Calculate maximum values for

(a) the acceleration of the box,
(b) its speed. [6]

(L)

In drawing free-body diagrams it is best to have each force arrow starting at the point of application of the force. When labelling the forces the use of push or pull may help too.

Solution 2.9

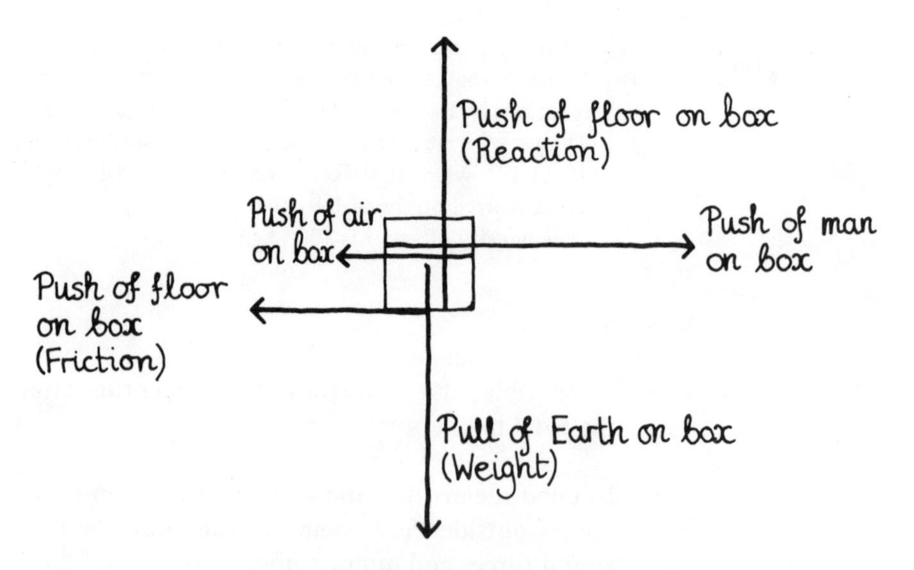

(a) The acceleration of the box is greatest when air resistance is zero. This is when the box starts to move.

Resultant force, $F_{res} = 4.5 \text{ N} - 1.5 \text{ N} = 3.0 \text{ N}$

By Newton's second law,

$$a = \frac{F_{res}}{m} = \frac{3.0 \text{ N}}{0.75 \text{ kg}} = 4.0 \text{ m s}^{-2}$$

(b) The maximum speed occurs when the box is in equilibrium.

The push of the man on the box = (air resistance) + (friction)

$$4.5 \text{ N} = kv^2 + 1.5 \text{ N}$$

Therefore

$$(6.0 \times 10^{-2} \text{ kg m}^{-1})v^2 = 3.0 \text{ N}$$

and hence $v = 7.1 \text{ m s}^{-1}$.

Example 2.10

According to Newton's third law, forces always exist as equal and opposite partners.
Name the partner of each of the forces X and Y shown in the diagram. [2]

(SEB)

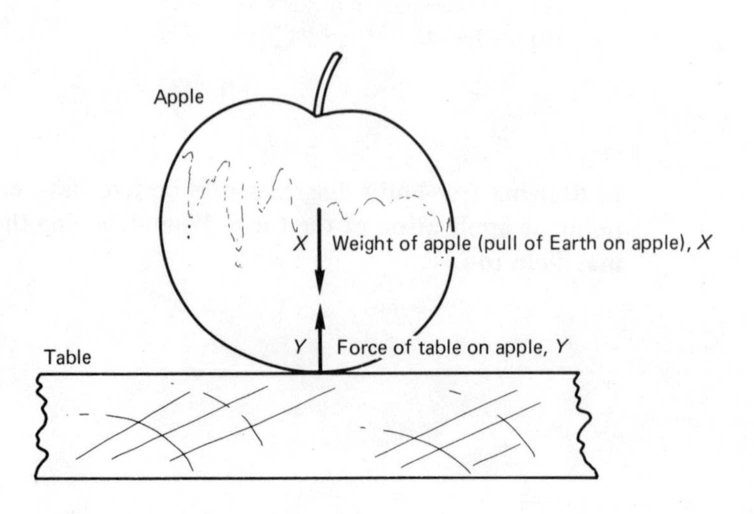

Solution 2.10

The pull of the Earth on the apple X is equal to the pull of the apple on the Earth.

The pull of the apple on the Earth may have little effect on the Earth, but the force nevertheless exists and is the same size as the weight of the apple.

The force of the table on the apple Y is equal to the force of the apple on the table.

Labelling forces by saying what body is exerting the force on what other body is the key to understanding Newton's third law.

Example 2.11

A boy stands still on ice skates, holding a ball. At time $t = t_1$ he throws the ball towards a wall, where it makes an elastic collision. He catches the ball as it bounces back towards him. Assuming that there is no friction between his skates and the ice, sketch a graph showing the way in which the velocity, v, of the boy varies with time t during these operations.

(NISEC)

Solution 2.11

When he throws the ball he gains a velocity v_1 away from the wall and continues at that velocity until he catches it again. As the ball is then moving away from the wall, the catch increases his velocity to about $2v_1$. A graph of his velocity against time would be as shown.

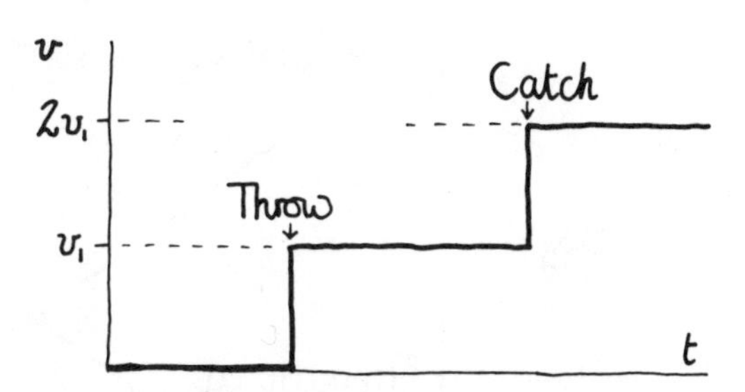

Try sketching a graph of the ball's velocity against time. Don't get caught out by thinking about conservation of momentum, it's not an isolated system!

Example 2.12

(a) Explain the relevance of the principle of conservation of energy to
  (i) a stone falling from rest near the Moon,
  (ii) a fireman sliding from rest down a vertical pole if he very soon reaches a steady speed owing to a frictional force opposing his motion.
  Sketch graphs, using one set of axes for the stone and another set for the fireman, showing how each form of energy you have considered varies with time. Mention the important features of the graphs. [10]

(b) A car of mass 800 kg is travelling at a constant speed of 30 m s^{-1} up a sloping straight road which rises 1.0 m for every 50 m along the road. When the brakes are applied and the engine stopped it takes 90 m for the car to come to a standstill. Using the principle of conservation of energy, calculate the frictional force on the car as it slows down and the acceleration. You may assume that both are steady while the car is slowing down. [10]

Solution 2.12

(a) (i) The stone is losing gravitational potential energy (g.p.e.) and gaining an equal amount of kinetic energy (k.e.).

(ii) The fireman loses g.p.e. of which some is changed into k.e. and some into heat as a result of the frictional force. Once he is going down with a steady speed the k.e. is constant and all the g.p.e. is being converted into heat.

(i)

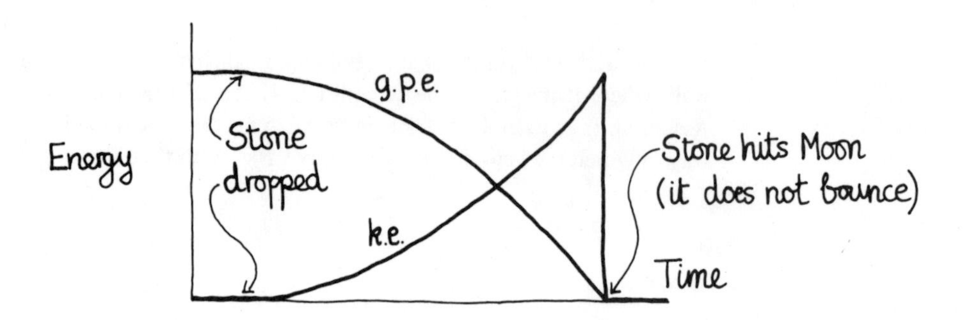

The gradients of the two graphs are equal and opposite at all times.

(ii)

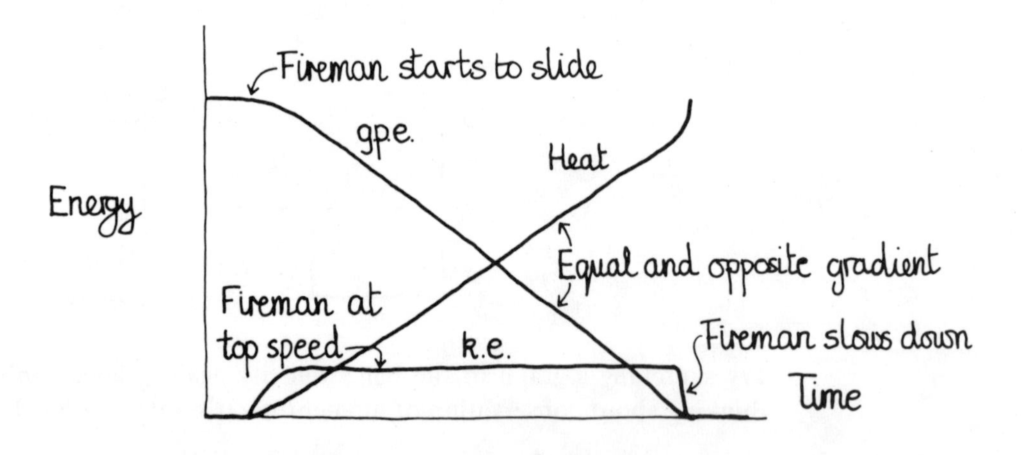

(b) The change in g.p.e. = (weight of car) × (height risen)
= (10 m s^{-2}) (800 kg) (90 m/50) = 14 400 J

The change in k.e. = $\frac{1}{2}mv^2$ = $\frac{1}{2}$(800 kg) (30 m s^{-1})^2 = 360 000 J

Some of the k.e. lost by the car is converted into g.p.e., but most is converted into heat as a result of the frictional force.

Energy converted into heat = (360 000 J) − (14 400 J) = 345 600 J

Frictional force = 345 600 J/90 m = 3840 N or 3.8 kN

The resultant force on the car parallel to the slope is the frictional force added to the component of the weight of the car in that direction.

Resultant force = (8000 N/50) + (3840 N) = 4000 N

This uses the fact that the slope is gentle and that the tangent of the angle to the horizontal and the sine of that angle are approximately equal to 1/50.

Acceleration in a direction up the slope = –4000 N/800 kg = –5.0 m s^{-2}

This could also be worked out using $v^2 - u^2 = 2as$ because the question states that a is constant.

Example 2.13

A ball of mass m falls vertically to the ground from a height h_1 and rebounds to a height h_2. The change in momentum of the ball on striking the ground is

A $mg(h_1 - h_2)$ **B** $m\sqrt{2g(h_1 - h_2)}$ **C** $m\sqrt{2g(h_1 + h_2)}$
D $m(\sqrt{2gh_1} - \sqrt{2gh_2})$ **E** $m(\sqrt{2gh_1} + \sqrt{2gh_2})$ (NISEC)

Solution 2.13

Suppose the ball hits the ground at a velocity v_1, using the principle of conservation of energy

$$mgh_1 = \tfrac{1}{2}mv_1^2$$

$$\Rightarrow \quad v_1 = \sqrt{2gh_1}$$

Similarly $v_2 = \sqrt{2gh_2}$, where v_2 is the velocity with which the ball leaves the ground.

The change of momentum of the ball (velocity is a vector) $= m\,(v_2 + v_1)$
$$= m(\sqrt{2gh_1} + \sqrt{2gh_2})$$

Answer E

Beware C which 'looks' a bit like it.

Example 2.14

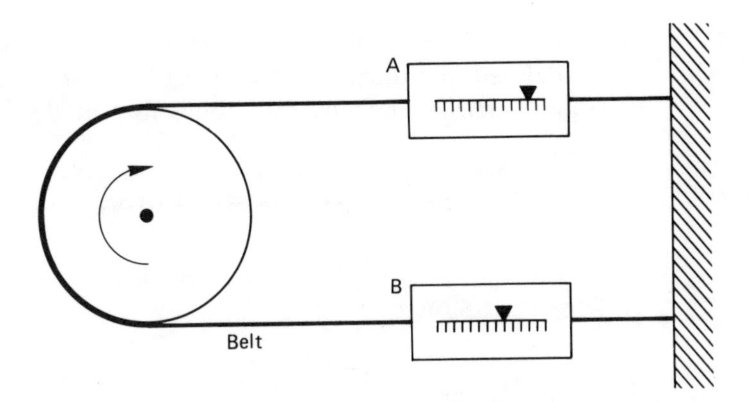

The diagram shows a light belt attached to two force meters and wrapped around a copper cylinder of mass 0.50 kg and radius 0.030 m. When the cylinder is rotated clockwise as shown, at three revolutions per second, the readings on the meters A and B are 100 N and 250 N respectively. At what rate is energy being dissipated because of friction between the belt and the cylinder? If the specific heat capacity of copper is 400 J kg^{-1} K^{-1} at what rate will the temperature of the cylinder start to rise when first rotated? If the rotation is continued why will the rate of rise of temperature probably decrease? [6]
(L)

Solution 2.14

The frictional force is the difference between the force meter readings:
$F = 250\text{ N} - 100\text{ N} = 150\text{ N}$
Power = $F \times$ (speed of surface) = (150 N) 2π (0.030 m) (3 s^{-1})
= 85 W

This formula is simply a consequence of the definition of work. If energy change = force × displacement, then energy change per second = force × displacement per second.

$\Delta Q = mc\Delta T$ where Q is the energy supplied to the bulk of the cylinder by heating and ΔT is the temperature rise.

Consider the first second

$\Delta Q = 85\text{ J} = (0.50\text{ kg})\,(400\text{ J kg}^{-1}\text{ K}^{-1})\,\Delta T$

$\Delta T = 0.43$ K so the rate of temperature rise is 0.43 K s^{-1}.

Once the cylinder is above room temperature it will start losing energy too and so the temperature will not rise as quickly.

See worked example 6.14.

Example 2.15

An alpha particle having a speed of 1.00×10^6 m s^{-1} collides with a stationary proton which gains an initial speed of 1.60×10^6 m s^{-1} in the direction in which the alpha particle was travelling.

What is the speed of the alpha particle immediately after the collision?

How much kinetic energy is gained by the proton in the collision?

It is known that this collision is perfectly elastic. Explain what this means.

(Mass of alpha particle = 6.64×10^{-27} kg. Mass of proton = 1.66×10^{-27} kg.) [5]

(L)

Solution 2.15

Since the numbers are tedious to write down it is best to start algebraically using letters defined in a diagram. Also using Mm for 10^6 m saves writing. The mass of the alpha particle is exactly four times that of the proton so this too saves work.

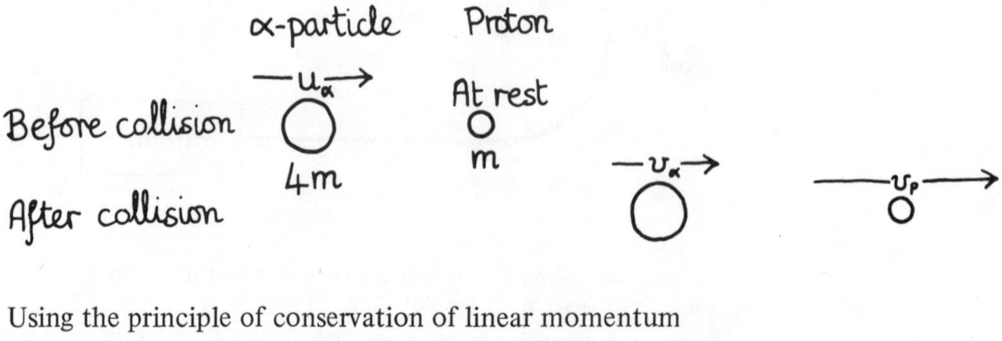

α-particle Proton
u_α At rest
Before collision
4m m
After collision v_α v_p

Using the principle of conservation of linear momentum

$$4mu_\alpha + 0 = 4mv_\alpha + mv_p$$

$$\Rightarrow 4u_\alpha = 4v_\alpha + v_p$$

$$v_\alpha = \frac{4u_\alpha - v_p}{4} = \frac{4\,(1.00\text{ Mm s}^{-1}) - (1.60\text{ Mm s}^{-1})}{4} = 0.60\text{ Mm s}^{-1}$$

k.e. of proton after collision $= \frac{1}{2}mv^2$
$= 0.5\ (1.66 \times 10^{-27}\ \text{kg})\ (1.60 \times 10^6\ \text{m s}^{-1})^2$
$= 2.12 \times 10^{-15}\ \text{J}$

Elastic means that the total k.e. after the collision is the same as the total k.e. before.

Example 2.16

(a) An isolated body of mass m, initially at rest, is acted upon by a constant force F. Derive an expression for the distance s travelled from the rest position in time t.
Write down an expression for the work done on the body by the force F and show that this work is equal to $\frac{1}{2}mv^2$, where v is the final velocity. [4]

(b) Define linear momentum and state the law concerning its conservation. How is linear momentum related to force? [4]

(c) Explain why, when catching a fast moving ball, the hands are drawn back while the ball is being brought to rest. Discuss whether your explanation has any bearing on the use of crushable boxes for packing eggs. [4]

(d) A particle, travelling along a straight path with velocity v, explodes into two equal parts. The forces due to the explosion act along the original direction of travel. The explosion causes the kinetic energy of the system to be doubled. Determine the subsequent velocities of the two segments. [6]

(AEB 1983)

Solution 2.16

(a) **A sketch showing what is going on is well worth the time and effort.**

The body accelerates uniformally with acceleration, $a = \dfrac{F}{m}$

The distance travelled from rest, $s = \frac{1}{2}at^2 = \dfrac{F}{2m}t^2$

The final velocity, $v = at$

The work done $= Fs = (ma)(\frac{1}{2}at^2)$

$= \frac{1}{2}m(at)^2 = \frac{1}{2}mv^2$

(b) Linear momentum = (mass) (velocity) of a body.
If a constant resultant force F acts on a body for a time t then

Ft = change of momentum of body

The change of momentum is in the same direction as the force.

(c) A fast moving ball has a certain momentum, mv, all of which is lost as it comes to rest. The slowing force F must therefore act for a time t where

$Ft = mv$

so the longer t the smaller F and the long t is achieved by drawing back the hands. Exactly the same argument applies to egg boxes; if an egg box is dropped then t is increased by the box crushing, hence the decelerating force F is smaller.

(d) **It often pays to avoid fractions; hence 4*m* not *m*.**

Linear momentum is conserved

$$4mv = 2mv_1 + 2mv_2$$

$$\Rightarrow \quad 2v = v_1 + v_2$$

Kinetic energy is doubled

$$\tfrac{1}{2}(2m)v_1^2 + \tfrac{1}{2}(2m)v_2^2 = 2 \times \tfrac{1}{2}(4m)v^2$$

$$\Rightarrow \quad v_1^2 + v_2^2 = 4v^2$$

$$= (2v)^2$$

$\therefore$ from above $v_1^2 + v_2^2 = (v_1 + v_2)^2$

i.e. $2v_1 v_2 = 0$

So that v_1 must be zero and hence $v_2 = 2v$.

Check back to the momentum and energy statements to see a slip has not occurred.

Example 2.17

(a) Write down an expression for the force required to maintain the motion of mass m moving with constant speed v in a circle of radius R. In which direction does the force act?

(b) The diagram shows a toy runway. After release from a point such as X, a small model car runs down the slope, 'loops the loop', and travels on towards Z. The radius of the loop is 0.25 m.

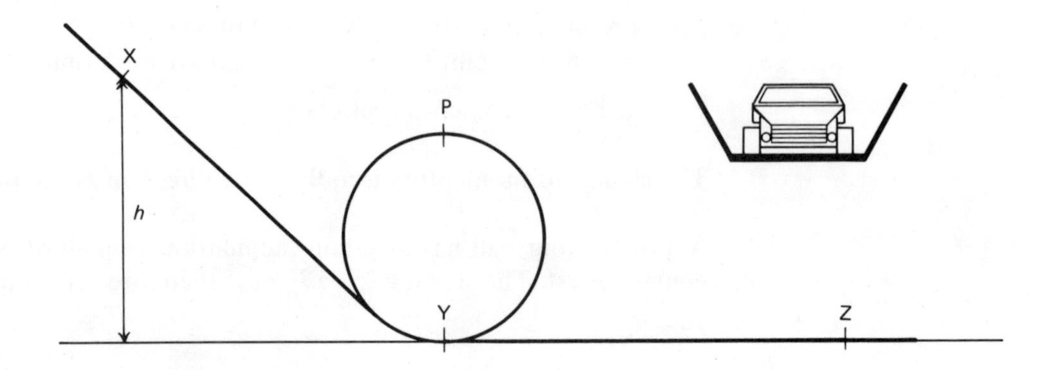

(i) Ignoring the effect of friction outline the energy changes as the model moves from X to Z.

(ii) What is the minimum speed with which the car must pass point P at the top of the loop if it is to remain in contact with the runway?

(iii) What is the minimum value of h which allows the speed calculated in (ii) to be achieved?

The effect of friction can be ignored. Assume that the acceleration of free fall is 10 m s^{-2}. [8]

(AEB 1984)

Solution 2.17

(a) The acceleration of a body moving in a circle is $\frac{v^2}{R}$

By Newton's second law the force required = $\frac{mv^2}{R}$

It is towards the centre of the circle.

(b) (i) From X to Y the car loses gravitational potential energy (g.p.e.) and gains kinetic energy (k.e.). From Y to P it loses some of its k.e. and gains g.p.e. From P to Y it loses the g.p.e. it gained and regains its original k.e. at Y. From Y to Z its k.e. and g.p.e. are constant.

(ii) At the top of the loop it is the weight of the car, mg, and the push of the track on the car, P, that keeps it moving in a circle.

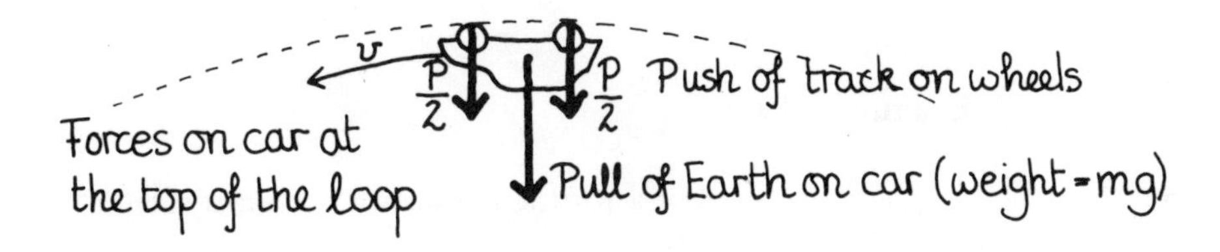

Using Newton's second law

$$\frac{mv^2}{R} = P + mg$$

As v gets less P tends to zero until

$$\frac{mv^2}{R} = mg$$

$v^2 = Rg = (0.25 \text{ m})(10 \text{ m s}^{-2})$, hence $v = 1.6 \text{ m s}^{-1}$.

(iii) The k.e. of the car at P is $\frac{1}{2}mv^2 = \frac{1}{2}mRg$

$(h - 2R)$ is the height difference between X and P so the g.p.e. change in going from X to P is $(h - 2R)mg$.

If friction is ignored,

k.e. change = g.p.e. change

$$\tfrac{1}{2}mRg = (h - 2R)mg$$

$$\Rightarrow \tfrac{1}{2}R = (h - 2R),$$

hence $h = 2.5R = 2.5\ (0.25 \text{ m}) = 0.63 \text{ m}$

Note that h is independent of the mass m of the car.

Example 2.18

An astronaut in an artificial satellite orbiting the Earth at a steady speed can be regarded as 'weightless' because:

A the gravitational force acting on him is zero.
B his gravitational potential remains constant.
C the centripetal force experienced by him is zero.
D his acceleration is zero.
E his acceleration is the same as that of the satellite.

Solution 2.18

Look at the five possibilities in turn.

A The gravitational field of the Earth extends a very long way. In any case most manned orbiters are close to the Earth and the pull of the Earth on the astronaut there is only a little less than that at the surface.

B The gravitational potential does remain constant, but it does at the Earth's surface too. This is not a reason.

C If something is moving in a circle it must by the very definition of acceleration be accelerating. A centripetal force is required to cause the continuous change in velocity. Though the speed is the same the direction of the velocity must be changing.

D See **C**

E CORRECT. Because both man and craft are falling freely with the same acceleration the craft exerts no supporting force on the man. It is the push of the floor on the man that gives him the sensation of weight and if this supporting force is missing he feels weightless and since he does not accelerate towards the floor he seems 'weightless'.

Example 2.19

(a) Explain why a particle moving with constant speed along a circular path has radial acceleration.

The value of such an acceleration is given by the expression $\frac{v^2}{r}$, where v is the speed and r is the radius of the path.

Show that this expression is dimensionally correct. [6]

(b) Explain, with the aid of clear diagrams, the following.

(i) A mass attached to a string rotating at a constant speed in a horizontal circle will fly off at a tangent if the string breaks.

(ii) A cosmonaut in a satellite which is in a free circular orbit around the Earth experiences the sensation of weightlessness even though he is influenced by the gravitational field of the Earth. [7]

(c) A pilot 'banks' the wings of his aircraft so as to travel at a speed of 360 km h^{-1} in a horizontal circular path of radius 5.0 km. At what angle should he bank his aircraft in order to do this? [5]

(L)

Solution 2.19

(a) Acceleration is defined as the rate of change of velocity. Velocity is a vector quantity, so when a body is moving in a circle at constant speed its velocity is continuously changing because its direction is changing.

The velocity is changing towards the centre of the circle so the acceleration is radial.

Dimensions of v are LT^{-1}.
Dimensions of r are L.
Dimensions of acceleration are LT^{-2}.

Dimensions of $\frac{v^2}{r}$ are $\frac{(LT^{-1})^2}{L} = LT^{-2}$, those of acceleration.

(b) (i) If the string breaks the mass will experience no further horizontal forces and will continue in a straight line.

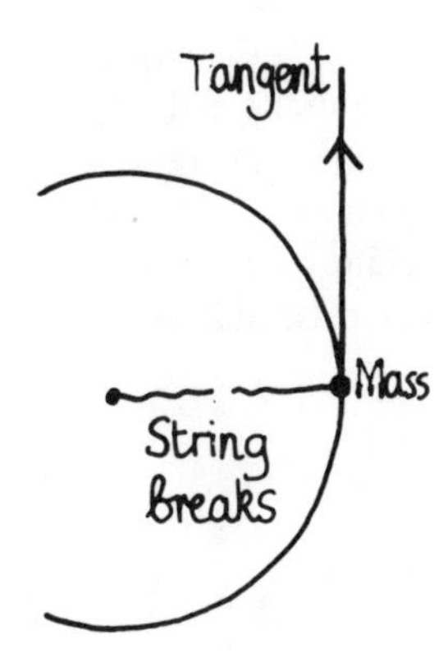

(ii) The sensation of weight is caused by the push of the floor on you. In orbit the cosmonaut and the floor of the spacecraft both have the same acceleration towards the centre of the Earth. The floor therefore exerts no supporting force on the cosmonaut.

(c)

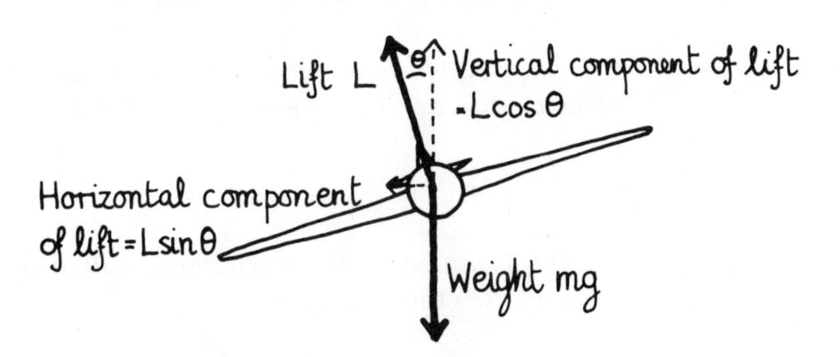

The vertical force on the plane = $L \cos \theta - mg = 0$

The horizontal force = $L \sin \theta = \frac{mv^2}{r}$ by Newton's second law.

$\sin \theta = \frac{mv^2}{rL}$ and $\cos \theta = \frac{mg}{L}$ so $\tan \theta = \frac{v^2}{rg}$ since m and L cancel.

$360 \text{ km h}^{-1} = 100 \text{ m s}^{-1}$

$\tan \theta = \frac{(100 \text{ m s}^{-1})^2}{(5.0 \times 10^3 \text{ m})(9.8 \text{ m s}^{-2})}$ so $\theta = 11.5°$

A common slip is to forget to convert speeds in km h^{-1} into m s^{-1} or to say there are 60 s in 1 h!

Example 2.20

(a) State Newton's law of gravitation and explain how this law was established. [3, 5]

(b) Use Newton's law to deduce expressions for:

(i) the period T of a satellite in circular orbit of radius r about the Earth in terms of the mass m_E of the Earth and the gravitational constant G. [3]

(ii) the gravitational field strength g at this orbit in terms of the orbital radius r, the gravitational field strength g_0 at the Earth's surface, and the radius r_E of the Earth (assumed to be a uniform sphere). [4]

(c) A satellite of mass 600 kg is in a circular orbit at a height 2000 km above the Earth's surface. Take the radius of the Earth to be 6400 km and the value of g_0 to be 10 N kg^{-1}. Calculate the satellite's:

(i) orbital speed; [5]

(ii) kinetic energy; [2]

(iii) gravitational potential energy. [5]

(d) Explain why any resistance to the forward motion of an artificial satellite results in an increase in its speed. [3]

(OLE)

Solution 2.20

(a) All bodies attract each other with a force proportional to the product of their masses and inversely proportional to the square of their distance apart.

Kepler made three general observations or laws about the motion of the planets around the Sun. He used Tycho Brahe's data. Newton used his three laws of motion and that of gravitation to explain Kepler's laws. Later the existence of Neptune and Pluto were predicted by applying the law of gravitation to the slight irregularities in the orbits of the inner planets.

It is not possible to do laboratory experiments to test the law to a greater accuracy than do astronomical observations.

(b) (i) The centripetal acceleration $r\omega^2 = r\left(\frac{2\pi}{T}\right)^2$

This is caused by the gravitational pull of the Earth on the satellite and so by Newton's second law

$$mr\left(\frac{2\pi}{T}\right)^2 = \frac{Gm_E m}{r^2}$$

where m is the mass of the satellite. Rearranging

$$T = 2\pi\sqrt{\frac{r^3}{Gm_E}}$$

(ii) At Earth's surface

$$g_0 = \frac{Gm_E}{r_E{}^2} \quad \text{and so} \quad Gm_E = g_0 r_E{}^2$$

In orbit radius r,

$$g = \frac{Gm_E}{r^2} = \frac{g_0 r_E{}^2}{r^2}$$

This must be so for an inverse square law.

(c) (i) In a circular orbit the free fall acceleration is equal to the centripetal acceleration, $g = \dfrac{v^2}{r}$

$$\text{So } v = \sqrt{rg} = \sqrt{\frac{rg_0 r_E{}^2}{r^2}} = \sqrt{\frac{g_0 r_E{}^2}{r}}$$

$$= \sqrt{\frac{(10\ \text{N kg}^{-1})\,(6.4 \times 10^6\ \text{m})^2}{(8.4 \times 10^6\ \text{m})}} = 7.0 \times 10^3\ \text{m s}^{-1}$$

(ii) $\text{k.e.} = \frac{1}{2}mv^2 = 0.5\ (600\ \text{kg})\ (7.0 \times 10^3\ \text{m s}^{-1})^2$

$$= 1.5 \times 10^{10}\ \text{J}$$

(iii) $$\text{g.p.e.} = \frac{-Gm_E m}{r} = -\frac{g_0 r_E{}^2 m}{r}$$

$$= \frac{-(10\ \text{N kg}^{-1})\,(6.4 \times 10^6\ \text{m})^2\ (600\ \text{kg})}{(8.4 \times 10^6\ \text{m})}$$

$$= -2.9 \times 10^{10}\ \text{J}$$

For a circular orbit the k.e. is always half the magnitude of the g.p.e. If you vaguely remember this result it provides a quick check.

(d) If the satellite's forward motion is reduced it will go into a lower orbit, thus losing g.p.e. and gaining k.e. It speeds up as a result despite some energy turning into heat.

2.8 Questions

Question 2.1

A tennis ball is dropped from the hand, falls to the ground and bounces back at half the speed with which it hit the ground. Draw a velocity-time graph of its motion. Mark the point on the graph which corresponds to the ball hitting the ground.

Indicate how, from the graph,

(a) the distance the ball falls, and

(b) the distance the ball rises,

can be found. [5]

(L)

Question 2.2

A car of mass 800 kg is towing a trailer of mass 200 kg. Both are accelerating steadily at 0.50 m s^{-2}. They are getting faster. Assuming there are no frictional losses calculate

(a) the forward force of the road on the drive wheels of the car. [2]

(b) the power developed by the car 4.0 s after starting from rest. [2]

(c) the force exerted by the trailer on the car. Give the direction as well as the size of the force. [2]

Question 2.3

The diagram shows speed–time graphs for three spherical objects made of the same material but with different radii, a_1, a_2, a_3, falling through a column of liquid.

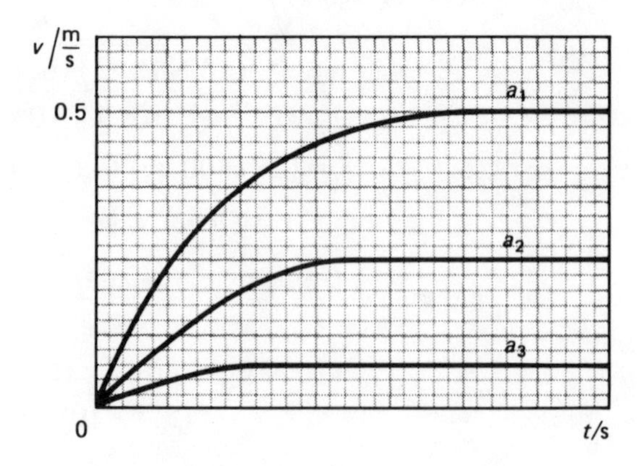

(a) Explain why the speeds tend to a constant value and why these values differ.
(b) If the constant speed of a falling sphere is proportional to its radius squared, find a_2 given that a_1 = 2.0 mm. [6]

(L)

Question 2.4

A car accelerates uniformly from rest to 20 m s^{-1} in 10 s, maintains this speed for 40 s, then is uniformly brought to rest over a further 5 s.

Sketch the velocity/time graph for the whole journey. Calculate
(a) the total distance covered,
(b) the mean speed. [7]

Question 2.5

What is an elastic collision? [2]

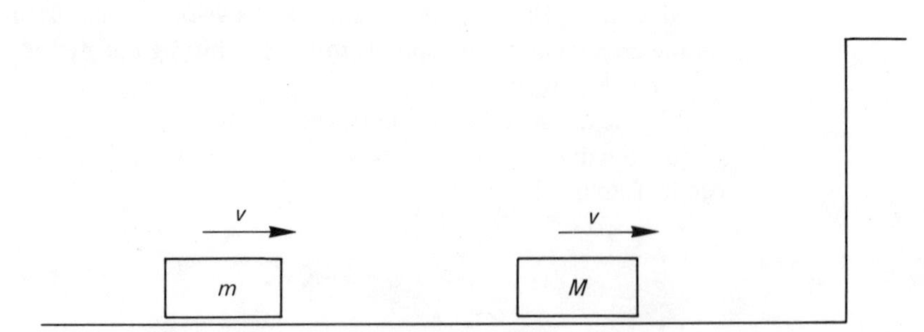

Two discs of masses m and M are sliding on a friction-free surface with the same speed v along a line at right angles to a fixed wall, as illustrated in the figure. M collides elastically with the wall and rebounds to strike m. Given that this collision is also elastic and the mass M is very much greater than m ($M \gg m$), find
(i) the final velocity of m; [5]
(ii) the change in momentum and kinetic energy of m as a result of these collisions. [2, 2]

(OLE)

Question 2.6

A spacecraft engine ejects high speed exhaust gases.

(a) Explain how this propels the spacecraft forward by making clear reference to a principle of physics.

At a particular time a rocket ejects 2500 kg of burnt fuel per second at a speed of 2000 m s^{-1} relative to the spacecraft.

(b) Calculate the force exerted on the spacecraft.

(c) This force remains constant for a time but the acceleration of the spacecraft changes. Explain why this is so. [7]

Question 2.7

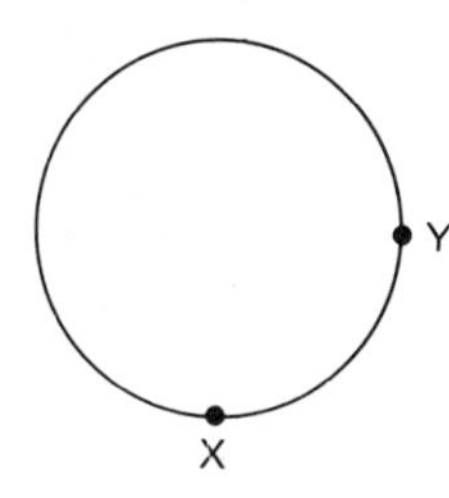

A particle moves round the circumference of a circle with constant speed in an anticlockwise direction. Which ONE of the following represents the direction of the change of velocity when the particle moves from X to Y? (NISEC)

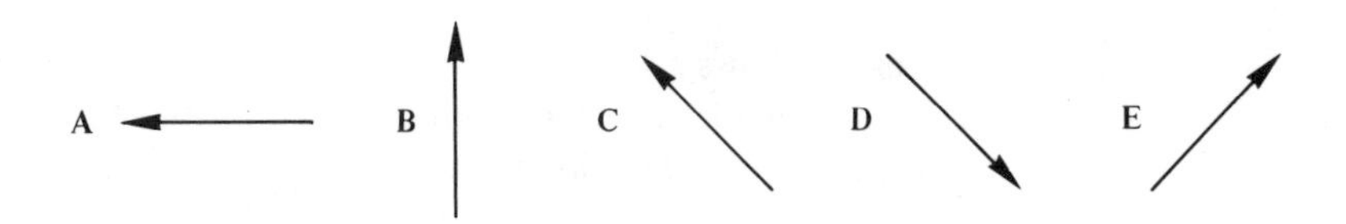

Question 2.8

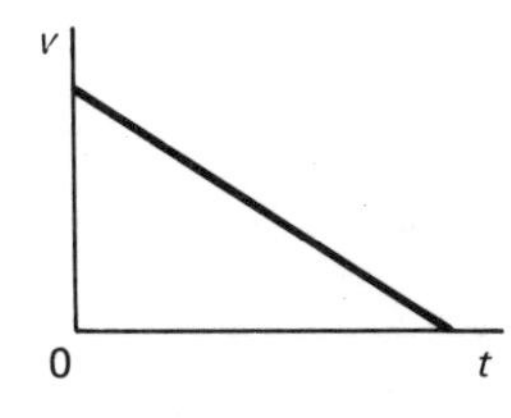

A bullet enters a fixed wooden block at time $t = 0$ and is brought to rest in time t_1. The graph shows the way in which the speed v of the bullet depends on time t. Sketch a graph to show how the force F exerted by the block on the bullet varies with time t. (NISEC)

Question 2.9

(a) What physical events occur at a molecular level during the non-elastic collision of two solid bodies? [3]

(b) Outline an experiment that could be carried out with laboratory equipment to determine the fraction of the available kinetic energy dissipated during the collision of two solid bodies. Describe the apparatus used, the procedure, and how the result would be obtained. [8]

(c) A car travelling at a steady speed of 30 m s^{-1} is acted on by air resistance which may be regarded as equivalent to that experienced by a flat surface of area 0.50 m^2 held with its plane normal to the direction of motion. As a simplification, the incident air is taken as stationary before impact with the car and assumed to accelerate to the speed of the car on impact. All other aerodynamic effects are neglected.
(Take the density of air to be 1.2 kg m^{-3}.)
Calculate:

(i) the mass of air striking the surface per second; [3]

(ii) the force acting on the surface; [3]

(iii) the work done against this force during a 100 km journey made at a constant speed of 30 m s^{-1}; [3]

(iv) the power exerted to sustain this constant speed. [3]

In order to overcome this air resistance, 15% of the mechanical output of the car engine is used. What is the total petrol used during the 100 km journey if the engine produces 40 MJ for each litre of petrol consumed? [4]

(v) How is the residue of an engine's mechanical output utilised after air resistance has been accounted for? [3]

(OLE)

Question 2.10

A small bob is suspended from a fixed point by a string 0.50 m long. It is made to rotate in a horizontal circle of radius 0.30 m, the centre of this circle being vertically below the point of support. Draw a diagram showing the forces acting on the bob, as seen by an external observer, when it is rotating. Calculate the period of rotation. [7]

(AEB 1983)

Question 2.11

The diagram shows a section through part of the Earth's gravitational field. The dashed lines, called equipotential lines, are perpendicular to the field lines. The gravitational potential, relative to a zero value at infinity, is shown at each of the three equipotential lines.

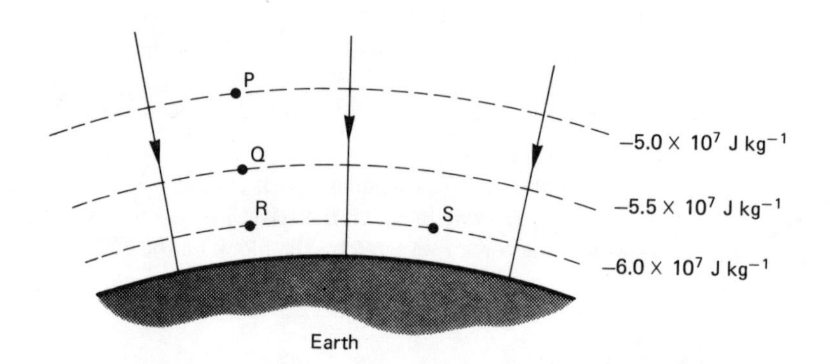

(a) (i) Is the gravitational potential bigger at P or at R? Explain your reasoning.
(ii) Calculate the change in gravitational potential energy of a spacecraft of mass 4000 kg moving from P to S. [5]

(b) The equipotential lines, which are given every 0.5×10^7 J kg^{-1}, are not equally spaced, i.e. the distance from P to Q is greater than the distance from Q to R. Calculate the distances PQ and QR given that, for the Earth, the mass of the Earth times the constant of gravitation, $m_E G = 4.0 \times 10^{14}$ N m^2 kg^{-1}. Hence explain why the equipotential lines are not equally spaced. [5]

Question 2.12

(a) State Newton's universal law of gravitation. What are the units of G, the constant of gravitation? [4]

(b) Show that the free-fall acceleration body at the Earth's surface, g_0, can be expressed as

$$g_0 = \frac{4}{3}\pi \rho G r_E$$

where ρ and r_E are the average density and the radius respectively of the Earth. [4]

(c) Describe a laboratory experiment to measure g_0. Explain how you would use your measurements to calculate the final result. [7]

(d) The Moon moves around the Earth in an orbit which is approximately circular and of radius 60 r_E, where r_E = 6400 km.

Taking g_0 to be 9.8 m s^{-2}:

(i) calculate the Moon's acceleration towards the Earth,

(ii) estimate the speed of the Moon relative to the Earth. [6]

Question 2.13

Explain how values of
g_0, the gravitational field strength at the Earth's surface,
G, the gravitational constant, and
r_E, the mean radius of the Earth
can be used to estimate the mean density of the Earth ρ_E. [5]

2.9 Answers to Questions

2.1 Remember a freely 'falling' object has exactly the same acceleration all the time. Only during the bounce itself will the gradient of the graph be different.

The distances of falling and rising could be thought of as positive and negative displacements. Displacement is to distance as velocity is to speed: one is a vector quantity and the other a scalar.

2.2 (a) Treat the car and trailer as one since only the car has drive wheels.
Answer 500 N

(b) Find the speed of the car and trailer and then calculate the power.
Answer 1000 W

(c) Find the force exerted by the car on the trailer and then use Newton's third law.
Answer 100 N opposite to the direction of the motion of the car.

2.3 (a) Sketch a force diagram for a drop. There will be an upthrust force caused by the displaced liquid.

(b) $v = ka^2$; work from graph data.

Beware the nasty vertical scale to the graph.

2.4 (a) 950 m, (b) 17 m s^{-1} (17.3).

2.5 See worked example 2.8.

(i) Draw a sketch showing m and M before and after their collision. As $M \gg m$ the velocity of m reverses and rises almost to $3v$ while that of M is almost unchanged.

Think of travelling with *M*, a truck, and meeting *m*, a bouncy ball thrown at the truck from in front. You would see it approach you at 2*v* and *rebound* from you at just less than 2*v*.

(ii) Show the new velocities on your diagram.
Momentum change of $m = 4mv$ to left.
Change of k.e. of $m = 4mv^2$.

2.6 (a) See worked example 2.7.
(b) Force is rate of change of momentum.
Answer 5.0×10^6 N
(c) Think of the mass of the spacecraft.

2.7 Sketch the circle and add velocity arrows at X and Y.
Find the vector velocity you need to add to the velocity at X to get the velocity at Y. A very basic question.
Answer C

2.8 The key is to consider the acceleration of the bullet as it enters the block and to go from there to F.

2.9 (a) See worked example 6.3.
(b) Look up a momentum conservation experiment in your textbook. From the collected data, you can calculate the k.e. as well as the momentum.

The word outline tells you that a fully labelled diagram will do for a description of the apparatus. This plus a list of measured quantities and the calculation is all you need.

(c) (i) and (ii) See worked example 2.8. (i) 18 kg s^{-1} (ii) 540 N
(iii) and (iv) Use the two equations for power given on page 14. (iii) 54 MJ (iv) 16 kW.
(v) 15% of the energy derived from the petrol = 54 MJ.
Hence find the petrol used at 40 MJ per litre.
Answer 9.0 litres

The last part says mechanical output and so is not about the engine's cooling system but asks where else energy is converted to internal energy.

2.10 There are two forces on the bob. The bob is not in equilibrium so apply Newton's second law using v^2/r for its centripetal acceleration in the horizontal plane.
The 3, 4, 5 triangle may help you in the calculation.
Answer 1.3 s

2.11 (a) (i) P (ii) 4×10^{10} J.

(b) Using $V = -\dfrac{Gm_E}{r}$ $\Rightarrow$ PQ = 730 km and QR = 610 km.

2.12 (a) Find this in your textbook or worked example 2.20. Beware of just quoting the formula. See page 13.
(b) The force given by Newton's law of gravitation causes the acceleration of the mass. Volume of a sphere = $\frac{4}{3}\pi r^3$.
(c) Use your textbook, but do not give an experiment for big G by mistake. Some of the experiments are so simple they could be done with things found in the home so you could try it. Most of the marks will be given for explaining how the result is calculated.
(d) (i) Use the inverse square law way in which the gravitational field strength falls off.
Answer 2.7×10^{-3} m s^{-2}

(ii) Use the centripetal acceleration formula and if you have time you can check your answer with the fact that the Moon goes round the Earth about every 28 days.
Answer 1.0 km s^{-1}

2.13 Show that $g_0 = G \dfrac{m_E}{r_E^2}$ and use $\rho_E = \dfrac{m_E}{V_E}$.

3 Oscillations and Waves

3.1 You Should Recognise

Quantity	*Symbol*	*Unit*	*Comment*
displacement	x, s	metre, m	a vector
period	T	second, s	pendulum of period 2.00 s has a length of 994 mm
frequency	f	hertz, Hz	equivalent to s^{-1}
amplitude	x_0	metre, m	
wavelength	λ	metre, m	optical wavelengths often in nm (10^{-9} m)
speed of wave	c, v	metre per second, $m\ s^{-1}$	speed of sound in air = $340\ m\ s^{-1}$
gravitational field strength	g	newton per kilogram, $N\ kg^{-1}$	at the Earth's surface g = $9.8\ N\ kg^{-1}$ or $9.8\ m\ s^{-2}$
mass per unit length	μ	kilogram per metre, $kg\ m^{-1}$	
spring constant	k	newton per metre, $N\ m^{-1}$	sometimes called spring stiffness
phase difference	ϕ	radian, rad	if wholly out of phase (in anti-phase) $\phi = \pi$ rad or 180°
s.h.m. constant	ω	per second, s^{-1}	sometimes called angular frequency
slit width	d	metre, m	
slit separation	s	metre, m	also grating spacing
order of interference	n	(no unit)	a whole number, also order of spectrum
focal length	f	metre, m	$1/f$ is called the power of the lens
speed of light *in vacuo*	c	metre per second, $m\ s^{-1}$	a constant = $3.00 \times 10^8\ m\ s^{-1}$
refractive index	n	(no unit)	for glass about 1.5
critical angle	θ_c	degree, °	less than 45° for glass

Not listed are quantities, such as mass and time, which occur in all chapters and which are given in the list on page 7.

3.2 You Should be Able to Use

- Simple harmonic motion:
  acceleration and displacement $a = -\omega^2 x$
  displacement and time $x = x_0 \sin \omega t$ or $x = x_0 \cos \omega t$
  maximum speed during oscillation $v_0 = \omega x_0 = 2\pi f x_0$

  frequency and period $f = \dfrac{1}{T}$, $\omega = 2\pi f$, $T = \dfrac{1}{f} = \dfrac{2\pi}{\omega}$
- Hooke's law $F = -kx$

- Time periods for simple harmonic oscillators:

simple pendulum $T = 2\pi \sqrt{\frac{l}{g}}$ for small (< about 10°) oscillations

mass on a spring $T = 2\pi \sqrt{\frac{m}{k}}$ for a spring of negligible mass

- Waves:

frequency and wavelength $c = f\lambda$

stationary waves distance between nodes = $\frac{\lambda}{2}$

mechanical wave speed on a stretched string, cord, etc. $c = \sqrt{\frac{T}{\mu}}$

electromagnetic waves in vacuum $c = \frac{1}{\sqrt{\epsilon_0 \mu_0}}$ (see page 120)

- Diffraction and interference:

diffraction at a slit $\sin \theta_m = \frac{\lambda}{d}$ for first minimum (normal incidence)

two-slit interference patterns fringe spacing = $\frac{\lambda}{s}$ × (distance slits-to-screen)

diffraction grating spectra at $n\lambda = s \sin \theta_n$ (normal incidence)

- Snell's law of refraction $\frac{\sin \theta_a}{\sin \theta}$ = constant for θ_a in air and θ in a medium

3.3 Simple Harmonic Motion

(a) Displacement–Time and Acceleration–Time Graphs for s.h.m.

See Fig. 3.1. If the oscillations are damped the period remains the same but the amplitude gradually reduces. The less the damping the greater the amplitude if the oscillator is forced to vibrate.

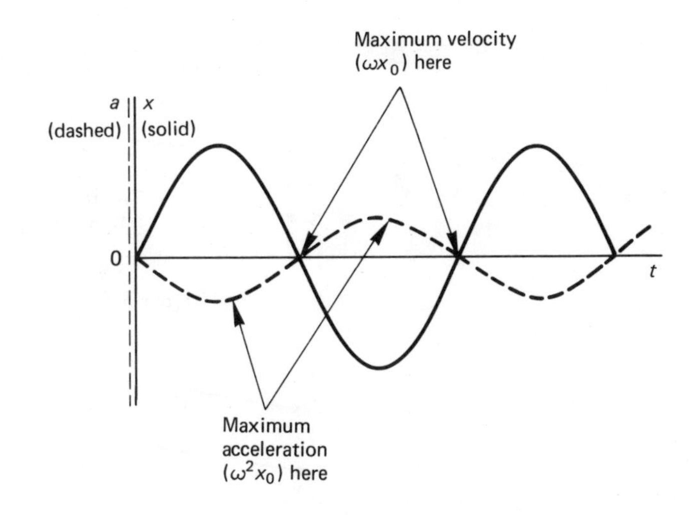

Figure 3.1

(b) Energy–Displacement Graphs for s.h.m.

See Fig. 3.2. The total mechanical energy is constant.

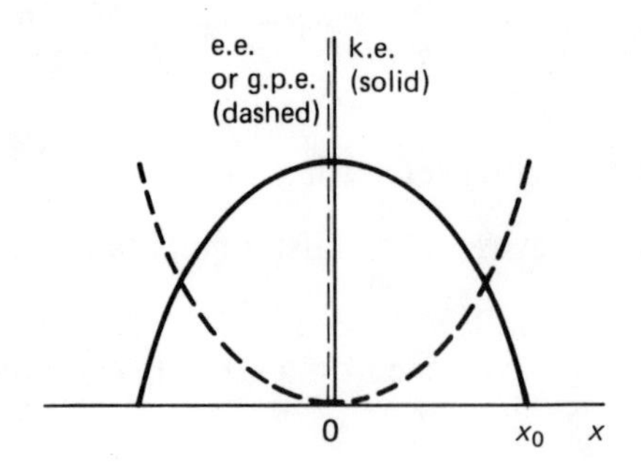

Figure 3.2

3.4 Waves

(a) Wave Properties

Waves transmit energy from place to place. For all waves $c = f\lambda$. The oscillations of points on a wave can be longitudinal as for sound waves, or transverse. Only transverse waves, e.g. electromagnetic waves, can be plane polarised. Points along a progressive sinusoidal wave are all oscillating at frequency f but the phase of the oscillation varies gradually along the wave.

(b) Interference Patterns

The principle of superposition, which all waves obey, states that the resulting displacement at a point P is the (vector) sum of the displacements caused by waves arriving at P from different sources, e.g. S_1 and S_2.

For two sources (or slits) that are in phase (see Figs 3.3 and 3.4), if

$S_2P - S_1P = n\lambda$ or $S_2N = n\lambda$, then there is a maximum at P,

but if

$S_2P - S_1P = n\lambda + \frac{1}{2}\lambda$ or $S_2N = n\lambda + \frac{1}{2}\lambda$, then there is a minimum at P.

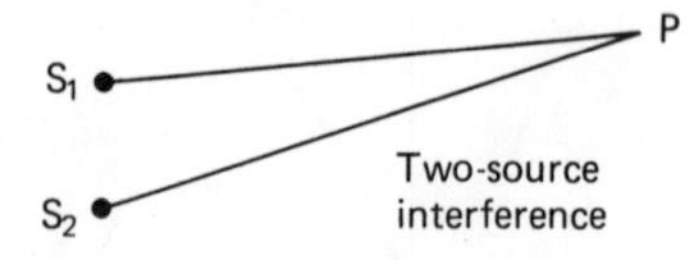

Figure 3.3

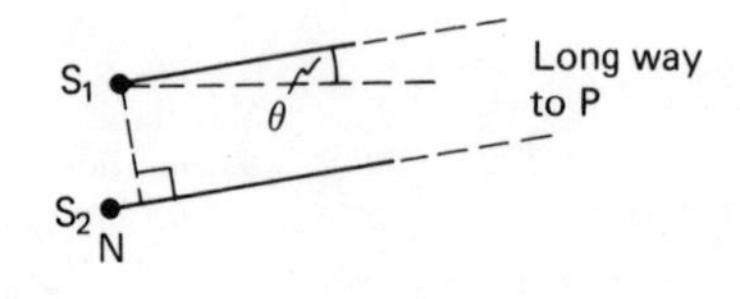

Figure 3.4

When S_1 and S_2 are light sources then, for them to be coherent (see example 3.10), they must both be derived from the same wavefront, e.g. by diffraction at two slits.

The distribution of energy in the resulting pattern of interference fringes depends on (i) the intensity of the diffracted light, which is determined by the ratio wavelength/slit width, and (ii) the superposition of the overlapping diffracted wavefronts, which produces maxima and minima within the diffraction envelope.

When many slits are used as in a diffraction grating, the maxima of the fringe pattern become very narrow so that different wavelengths (colours) in the incident light produce sharp lines and a spectrum of lines is produced on either side of the centre. The order of the spectrum is given by $S_2N = n\lambda = s \sin\theta$, where s is the slit separation.

(c) Stationary Waves

Stationary waves are the result of superposition between an incident wave and its reflection. For sound waves in air in a tube closed at one end (Figure 3.5) there is an antinode of displacement near to the open end and a node of displacement at the closed end. The oscillations of points between adjacent nodes are in phase; energy is not transmitted along stationary waves. For waves on strings, etc., which are fixed at both ends, there must be a node at each end as in Fig. 3.6. The different modes of vibration with which air tubes and stretched strings can oscillate are characterised by their resonant frequencies. If the lowest or fundamental frequency is f then higher frequencies, which are multiples of f, are called harmonics.

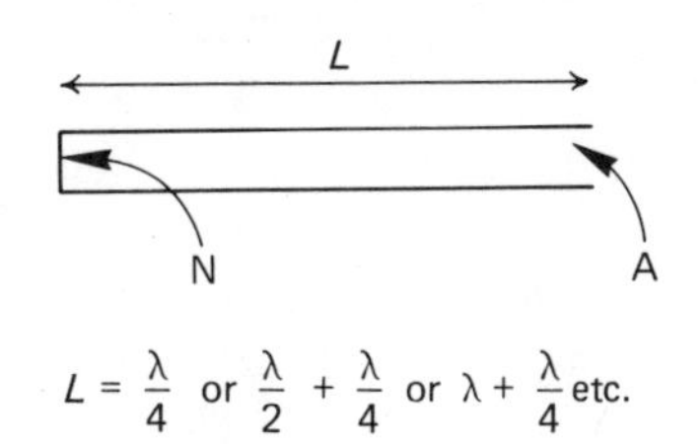

$$L = \frac{\lambda}{4} \text{ or } \frac{\lambda}{2} + \frac{\lambda}{4} \text{ or } \lambda + \frac{\lambda}{4} \text{ etc.}$$

Figure 3.5

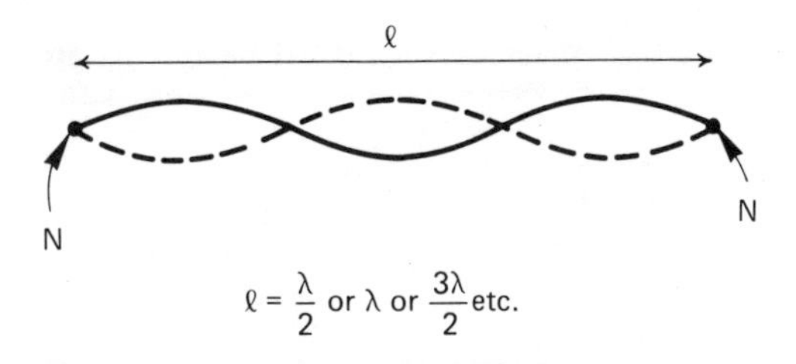

$$\ell = \frac{\lambda}{2} \text{ or } \lambda \text{ or } \frac{3\lambda}{2} \text{ etc.}$$

Figure 3.6

3.5 The Electromagnetic Spectrum

The main divisions of the electromagnetic spectrum are shown in Fig. 3.7 together with approximate wavelength ranges. Light is from about 4×10^{-7} m to about 7×10^{-7} m. Some regions overlap because waves produced in different ways are

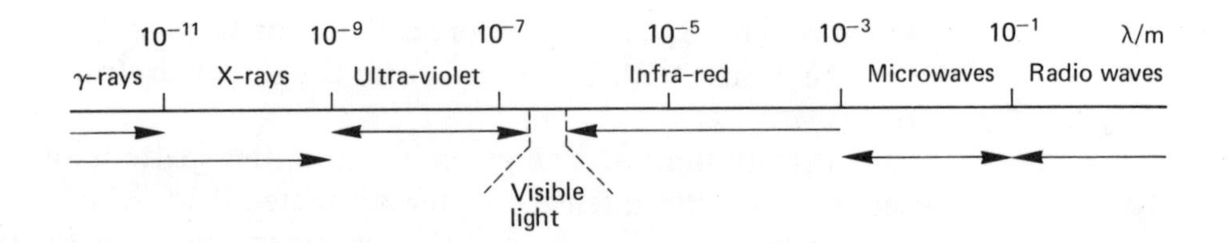

Figure 3.7

given different names, e.g. γ-rays come from the nucleus while X-rays can be produced by a beam of electrons striking a metal target.

3.6 Ray Optics

The paths taken by wave energy can be described by rays. Rays are perpendicular to wavefronts. If in Fig. 3.8 θ_a is 90°, then θ is called the critical angle θ_c, and $\sin \theta_c = 1/n$. When $\theta > \theta_c$ the wave is totally reflected.

When a converging lens of focal length f produces two real images of distant objects (one set of parallel rays might be from the top of the Moon, the other from the bottom) with an angular separation ϕ, then the images are $f\phi$ apart (ϕ must be in radians) – see Fig. 3.9.

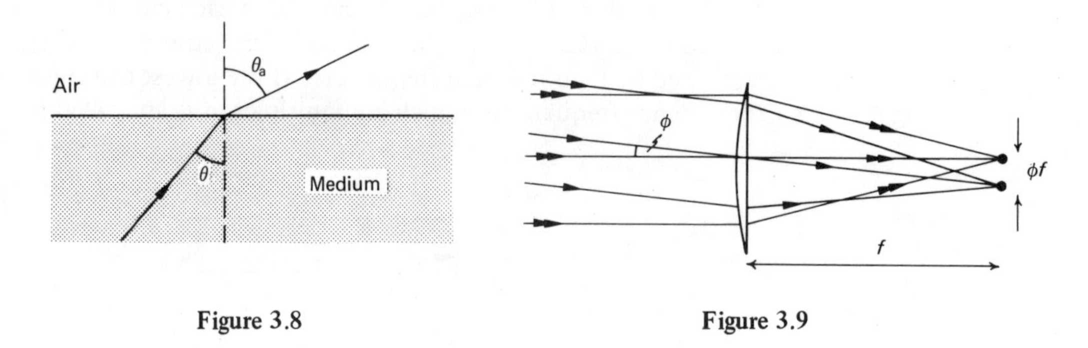

Figure 3.8 **Figure 3.9**

3.7 Worked Examples

Example 3.1

A 'wig-wag' machine or inertial balance is an example of a mechanical oscillating system.

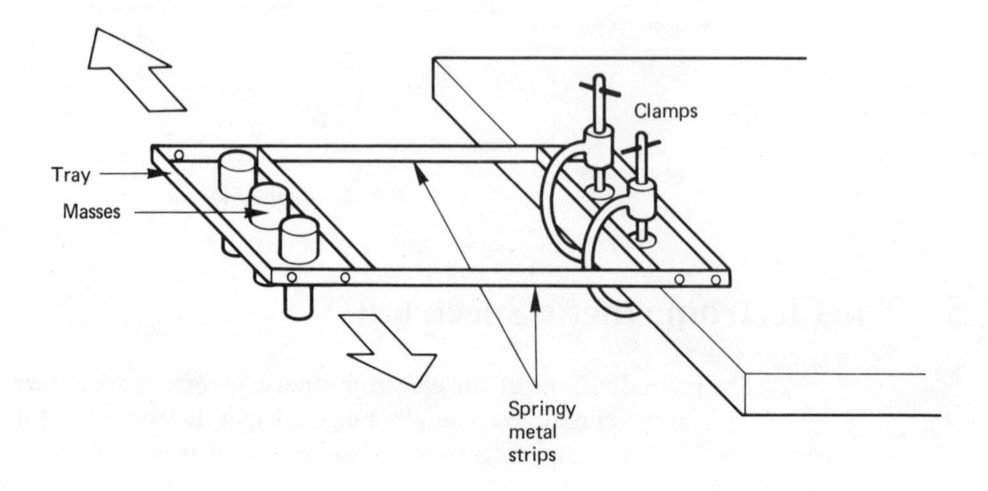

One, two, or three identical masses can be pegged into holes in the tray, providing units of mass for this experiment. The tray is pulled to one side and then released to set it oscillating.

The graph was obtained for such a device.

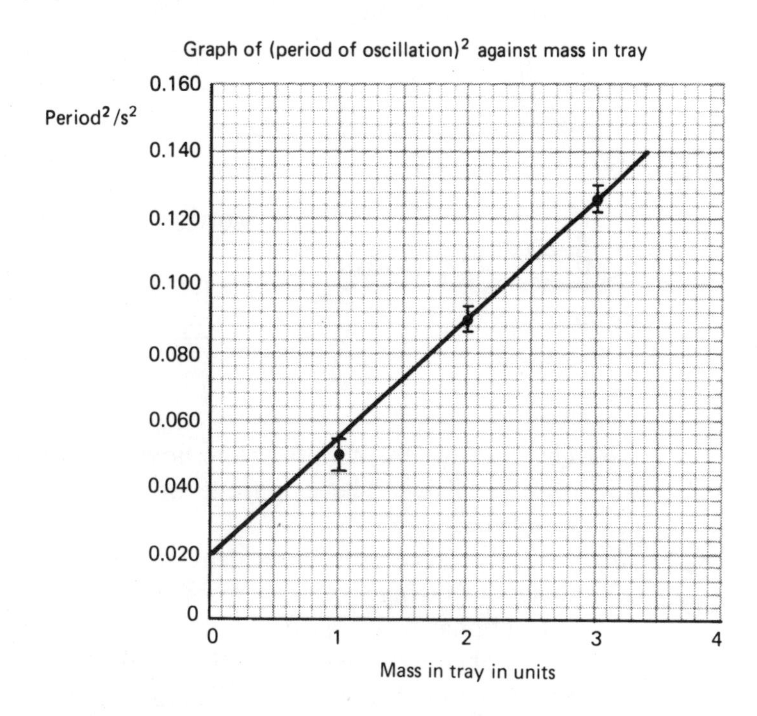

(a) Extend the graph and read off the mass of the empty tray.

(b) State the relationship between period and the total moving mass suggested by the graph. Justify your conclusion.

When the tray is set oscillating with only a 0.50 kg mass secured in it the period is measured to be 0.29 s.

(c) Use the graph to find the mass of one unit mass in kilograms. [5]

(SEB)

Solution 3.1

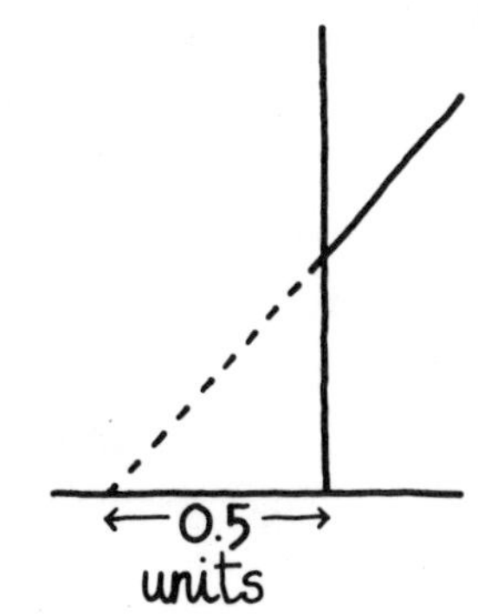

(a) The period would be zero if the total mass, including the tray, were zero. Using the graph to find where the line hits the x-axis: mass of tray is 0.5 units.

(b) The graph is straight when the (period)^2 is plotted and would go through the origin if total mass were plotted. Period is proportional to (total mass)$^{\frac{1}{2}}$.

(c) $(0.29\ \text{s})^2 = 0.084\ \text{s}^2$, from the graph this corresponds to about 1.8 units of mass.

So one unit $= \dfrac{0.50\ \text{kg}}{1.8} = 0.28\ \text{kg}$

Example 3.2

Oscillations occur in varied branches of physics and drawing analogies between different types of oscillation can aid understanding.

You are supplied with a 1 kg mass dynamics trolley, springs, a battery, a cathode ray oscilloscope, a 50 H inductor and a 10 mF capacitor. You also have the sort of equipment normally found in a physics laboratory.

(a) Draw diagrams to show how the equipment could be set up to demonstrate both mechanical and electrical oscillations. [6]

(b) Describe the electrical oscillations if a resistor is connected in series with the inductor. [3]

(c) What are the equivalents of inductance and resistance in the mechanical oscillations? [2]

(d) Would the e.m.f. of the battery used to start the oscillations affect the frequency of the electrical oscillations? [1]

Solution 3.2

(a)

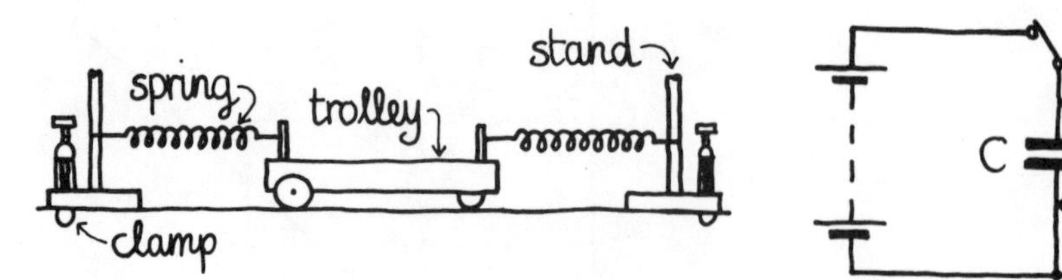

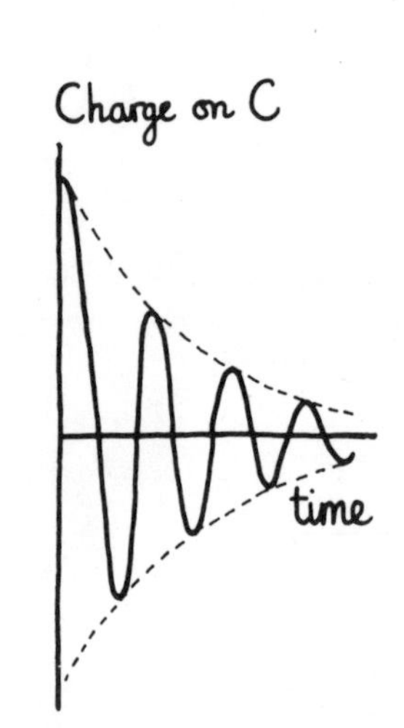

(b) The diagram on the left shows how the charge on C indicated by the c.r.o. trace oscillates. The amplitude of the oscillations decreases exponentially with time.

(c) The inductance is the equivalent of mass because energy is stored in it when charge is flowing just as the mass has k.e. when the mass is moving.

Frictional forces including drag are the equivalent of resistance in the electrical circuit. They cause the oscillations to die away too.

(d) No. The initial amplitude of simple harmonic oscillations does not affect the frequency.

Example 3.3

(a) State the conditions necessary for the motion of an oscillating body to be simple harmonic. Give one reason why the vertical oscillations of a body suspended from a spring may not satisfy these conditions.

(b) The vertical oscillations of a body on a spring are started by holding the body at a point where the spring is at its natural length and then releasing it.

State and explain briefly the effect of increasing the mass of the body on the value of each of the following quantities:

(i) the time period of the oscillation;

(ii) the amplitude of the oscillation;

(iii) the total energy of the oscillating system. [8]

(AEB 1984)

Solution 3.3

(a) If the body is to make simple harmonic oscillations when displaced from its equilibrium position, the size of the force on the body towards the equilibrium position must be proportional to the distance of the body from the equilibrium position. Using vectors one may say that the resultant force on the body is proportional to its displacement, but in the opposite direction.

One reason for a body hung on a spring not performing simple harmonic motion is the drag of the air on the body. This force is not proportional to its displacement and in fact depends on the speed of the body.

Another reason might be that the spring does not obey Hooke's law exactly. In particular in the case of large displacements the tension in the spring may no longer increase in exact proportion to its extension.

(b) (i) Increasing the mass of the body will increase the period of the oscillations. The tension in the spring will be the same for a particular extension and so if the mass is larger the acceleration will be less in accordance with Newton's second law. If the acceleration of the body is less then a complete oscillation will take longer to perform.

(ii) The equilibrium position will be lower for a larger mass since the spring will extend more before the upward pull of the spring on the body is equal to the downward pull of the Earth on the body. Amplitude is the maximum displacement from the equilibrium position, so if the top of the oscillation is in the same place, but the centre is lower, then the amplitude of the oscillation of the larger mass will be greater.

(iii) The total energy of the oscillating system will be equal to the maximum kinetic energy (k.e.) as the body passes through its equilibrium position. Consider two different masses on the same type of spring passing through the equilibrium position of the smaller mass. Both springs will store the same elastic energy, but because the larger mass has lost more gravitational potential energy it will have more k.e. than the smaller mass. And the larger mass has not even reached its maximum k.e. at its own equilibrium position. Hence the total energy of the oscillations of the larger mass must be larger.

Example 3.4

(a) One end of a spring is attached to a rigid support and a mass is hung on the other end. Show that the mass will oscillate with simple harmonic motion after it has been moved vertically from its equilibrium position and derive an expression for the period of the oscillation. You may assume that the spring obeys Hooke's law and that its mass is negligible. [6]

(b) A vehicle suspension system consists of springs together with shock absorbers to damp oscillations. The shock absorber consists of a piston in a cylinder filled with oil. A vehicle of mass 800 kg on which the shock absorbers are *not* functioning is pushed down 50 mm, released and allowed to oscillate. What is the period of the oscillation if the vehicle's body would have to have been lifted 100 mm before the wheels left the ground? You may ignore the mass of the wheels compared with the mass of the body. [4]

(c) The shock absorbers are now repaired. Sketch two graphs on the same axes to compare the oscillations of the vehicle before and after the repairs. That is with and without damping. [3]

(d) Discuss the energy changes which occur during the oscillations,
  (i) when the shock absorbers are not working,
  (ii) when the shock absorbers are working. [7]

Solution 3.4

(a)

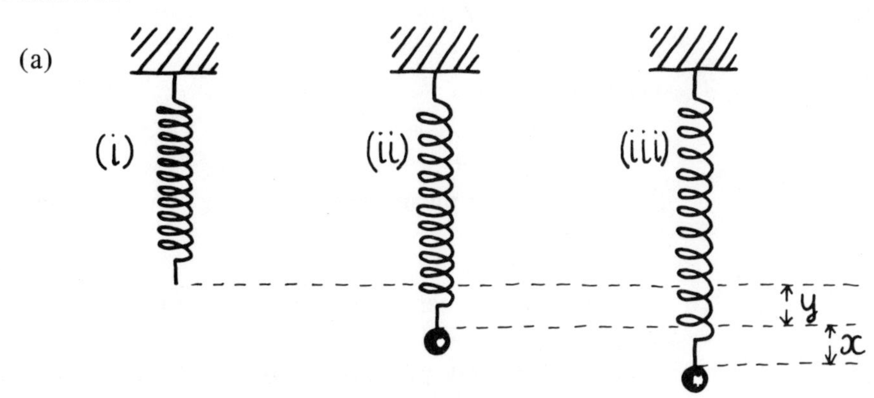

Hooke's law can be written as

Tension in spring = $k \times$ displacement

Diagram (ii) shows the body in equilibrium so its weight is equal to the pull upwards of the spring.

$$mg = ky \qquad (*)$$

Think of the mass in diagram (iii).

The accelerating downward force $= mg - k(x + y)$
$= mg - kx - ky$
$= -kx$ by equation (*)

The − sign means the force and hence the acceleration are in the opposite direction from the displacement.

Using Newton's second law $ma = -kx$

$$a = -kx/m$$

This meets the condition for s.h.m. that the acceleration towards the equilibrium position is proportional to the displacement from it.

The general formula for s.h.m. is $a = -\omega^2 x$, where $\omega = 2\pi f$.
In this case $\omega^2 = k/m$, so the period, $T = 2\pi\sqrt{m/k}$.

(b) The springs compress 0.10 m when a force of 8000 N is applied, because we know 1 kg weighs 10 N.

The spring constant, $k = \dfrac{8000 \text{ N}}{0.10 \text{ m}} = 80\,000 \text{ N m}^{-1}$

Period $T = 2\pi\sqrt{\dfrac{800 \text{ kg}}{80\,000 \text{ N m}^{-1}}} = 0.63$ s.

The 50 mm mentioned in the questions is irrelevant here, since the period of s.h.m. is independent of its amplitude. Mass and weight both occur here; be careful because there is scope for confusion.

(c)

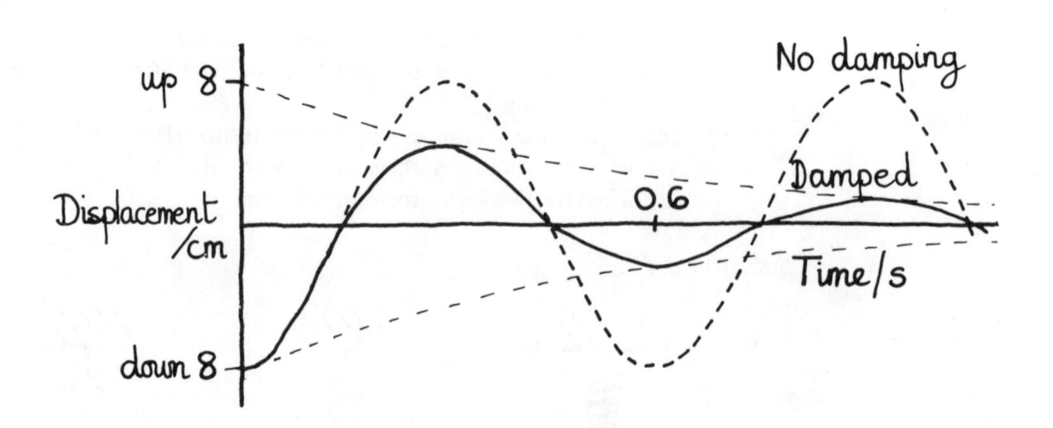

Sketch does not mean draw carelessly. Show as clearly as you can what is happening.

(d) (i) At first the car has gravitational potential energy, g.p.e., no kinetic energy, k.e., and a little energy stored in the springs, e.e.

As car body goes down g.p.e. is converted to k.e. and e.e.

At equilibrium k.e. reaches a maximum.

As body goes down further k.e. and g.p.e. are converted into e.e.

At the bottom there is a minimum of g.p.e., no k.e. and a maximum of e.e.

As the body rises again e.e. is converted back into g.p.e. and k.e.

The whole process is now in reverse and, once at the top again, will repeat.

(ii) With the shock absorbers working as soon as the car moves k.e. is converted into internal energy (i.e. heat) as a result of frictional forces.

The k.e. eventually gets less and the car body stops at its equilibrium position. The amount of damping decides how much the body oscillates before stopping.

Note how the use of defined abbreviations saves unnecessary writing and how a one idea per line format makes it easier to write energy 'stories'.

Example 3.5

The motion of the piston in a petrol engine is approximately simple harmonic. The frequency of the oscillation is 100 Hz and the amplitude 35 mm. Find

(a) the maximum acceleration of the piston

(b) the maximum velocity of the piston. [5]

Solution 3.5

(a) The basic formula for s.h.m. is $a = -\omega^2 x$ where $\omega = 2\pi f$. The maximum acceleration occurs when the displacement is equal to the amplitude x_0

$a_0 = -(2\pi f)^2 (x_0) = (2\pi \times 100 \text{ Hz})^2 (0.035 \text{ m}) = 1.4 \times 10^4 \text{ m s}^{-2}$

Note that when the piston is accelerating most it is temporarily stationary.

(b) The maximum velocity is $\omega x_0 = (2\pi \times 100 \text{ Hz}) (0.035 \text{ m})$
$= 22 \text{ m s}^{-1}$

If you remember that s.h.m. is like a side view of circular motion of radius x_0, then it becomes clear that the maximum speed of the s.h.m. is the same as the steady speed around the circle. The speed of a body moving in a circle is ωr.

Example 3.6

A light spring is loaded with a mass of 200 g and made to execute vertical oscillations. The diagram shows a force extension graph for the spring.

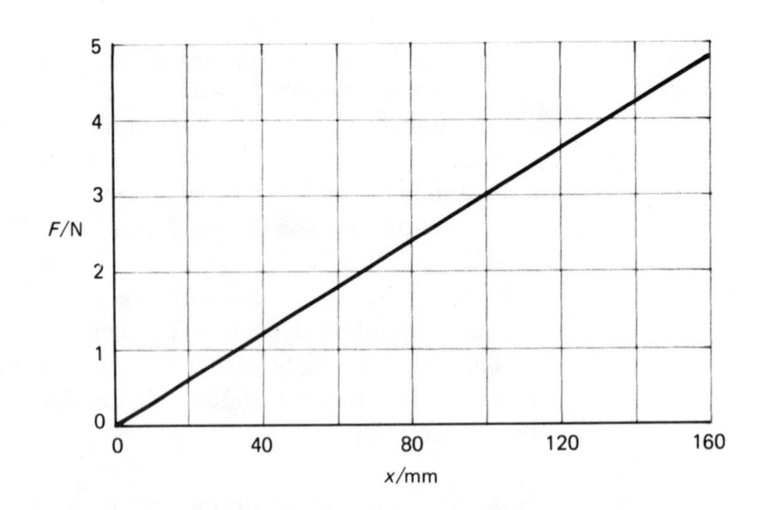

(a) Explain why the oscillations are likely to be simple harmonic.

(b) Find the slope, k, of the graph. Hence calculate the period of vertical oscillation of the 200 g mass.

(c) Calculate the energy stored in the spring when the mass is at its lowest point, if the amplitude of the oscillation is 20 mm. [7]

(L)

Solution 3.6

(a) The motion is simple harmonic motion because the force restoring the mass to its equilibrium position is always proportional to its distance from it. By Newton's second law the acceleration is also proportional to the distance: $a = -\omega^2 x$ ($\omega = 2\pi f$, the angular frequency).

(b) The graph is a straight line through the origin so

$$\text{slope} = \frac{4.8 \text{ N}}{160 \text{ mm}} = \frac{4.8 \text{ N}}{0.16 \text{ m}} = 30 \text{ N m}^{-1} \text{ to 2 sig. fig.}$$

$$\omega^2 = \frac{k}{m} = \frac{30 \text{ N m}^{-1}}{0.20 \text{ kg}} = 150 \text{ s}^{-2}$$

$$\therefore \text{ period} = \frac{2\pi}{\omega} = \frac{2\pi}{12.2 \text{ s}^{-1}} = 0.51 \text{ s}$$

(c) The mass is in equilibrium when its weight is equal to the pull of the spring upwards.

weight = (0.20 kg) (10 N kg^{-1}) = 2.0 N

From the graph the equilibrium position is therefore 67 mm. The lowest point is 67 mm + 20 mm = 87 mm down.

Energy stored is the area under the graph up to that point.

Area = $\frac{1}{2}$ x base x height = 0.5 (0.087 m) (2.6 N) = 0.11 J.

The 2.6 N was the height of the graph's triangular area.

Example 3.7

(a) Explain what is meant by ***resonance***.

Describe briefly three examples of resonance (each from a different field of physics) met in experimental work in a school laboratory. [8]

(b) Describe, illustrating the descriptions with suitable graphs:

(i) the mutual interaction of two similar vibrating systems which are not quite in resonance with each other;

(ii) the effects of progressively increased damping on resonant vibration. [8]

How would you investigate one of these effects? Describe the apparatus you would use and how the readings would be obtained. [6]

(c) A steel tube 3.0 m long, carrying a stream of gas in an industrial plant, is clamped at each end and subjected to transverse vibrations. The speed of transverse waves along the material of the tube is 400 m s^{-1}. Calculate the lowest frequency of standing wave that can form on the tube. [3]

The transverse vibrations of the tube cause pressure fluctuations in the gas column forming standing waves with vibration antinodes at either end. If the speed of sound in the gas is 300 m s^{-1}, calculate the lowest frequency at which the vibrations of the tube and of the gas column would resonate. [5]

(OLE)

Solution 3.7

(a) Resonance is the effect where a system free to oscillate is driven at a frequency very close to the frequency at which it would oscillate by itself if displaced. Even though the driving oscillation might not be powerful the energy stored in the system builds up and the amplitude of the oscillations becomes very large.

1. A taut wire between the poles of a permanent magnet carrying an alternating current will vibrate violently when the frequency of the a.c. is the same as the frequency of the note when the wire is 'twanged'.
2. A circuit with an inductor and capacitor connected in parallel to an alternating p.d., an *LC* circuit, has a large current in the inductor when the frequency of the small driving p.d. is equal to the resonant frequency of the circuit.
3. An electric motor mounted on spring supports may vibrate violently when the frequency of rotation of the motor is the same as the frequency at which the motor would naturally wobble on its supports.

2 is the principle of radio reception.

(b) (i) Energy will be passed back and forth between the two oscillators. The period of this exchange depends on the difference between the two natural frequencies.

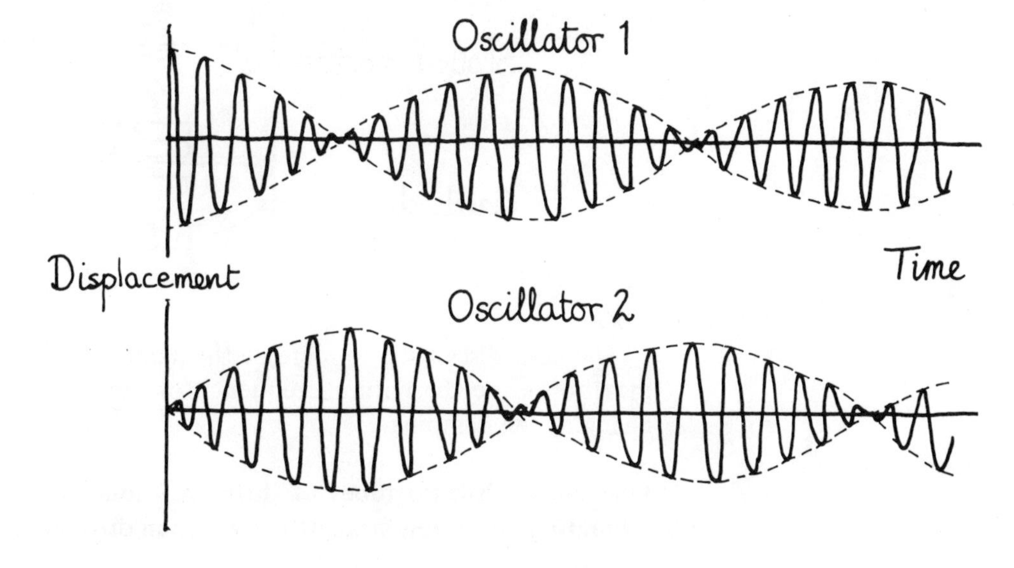

(ii)

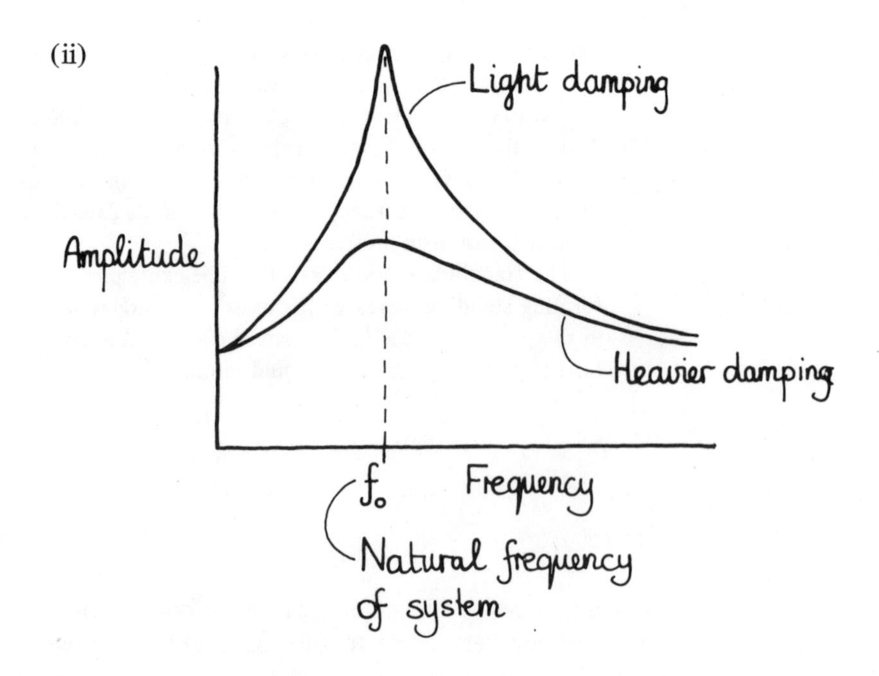

Damping reduces the resonant effect and makes the frequency at which it happens less sharp.

The equipment is set up as in the diagram below. The geared motor must be powerful enough to guarantee that its speed of rotation will be unaffected by whatever the spring is doing.

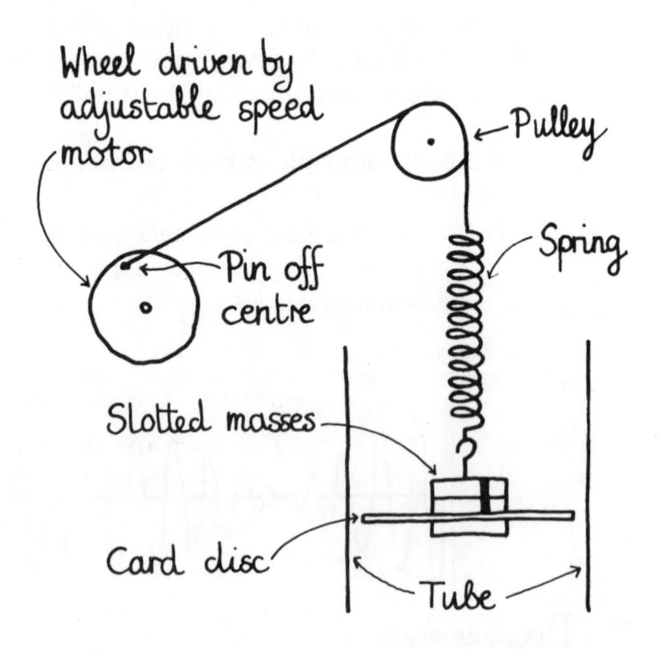

The card discs used to increase the damping should all have the same mass to ensure the natural frequency of oscillation is unaffected when changing the damping.

One could also use tubes of different cross-sections, but in a school laboratory this is less straightforward than different pieces of card.

The frequency of the motor's rotation is gradually increased from frequencies below that at which the spring oscillates by itself to frequencies above. The frequency of the motor at each voltage setting is measured either by a stroboscopic method or if the frequency is low enough by simply timing, say, 20 revolutions.

At each motor frequency the maximum amplitude of the oscillating mass is recorded and plotted on an amplitude against frequency graph.

Now the damping is changed and the process repeated to produce further graphs.

The experiment is simplest if the mass is large and the spring weak. This gives conveniently slow oscillations.

(c) The lowest frequency of the transverse wave will have a wavelength equal to half the length of the tube with a node at each end, i.e. $\lambda/2 = 3.0$ m and $\lambda = 6.0$ m

$$\Rightarrow \quad f = c/\lambda = 400 \text{ m s}^{-1}/6.0 \text{ m} = 67 \text{ Hz}$$

At resonance f will be the same for each standing wave, but since the sound wave has a lower speed, its wavelength will be shorter by a factor of $300 \text{ m s}^{-1}/400 \text{ m s}^{-1} = 3/4$.

This resonance will first occur when the transverse wave is at its 3rd harmonic and the longitudinal wave at its 4th harmonic.
Frequency = 3×67 Hz = 200 Hz

It is a good idea to check.
The wavelength of the sound = $300 \text{ m s}^{-1}/200 \text{ Hz} = 1.5$ m.
This fits into 3.0 m giving 4 half wavelengths.
Sketches of both waves as transverse waves may help too.

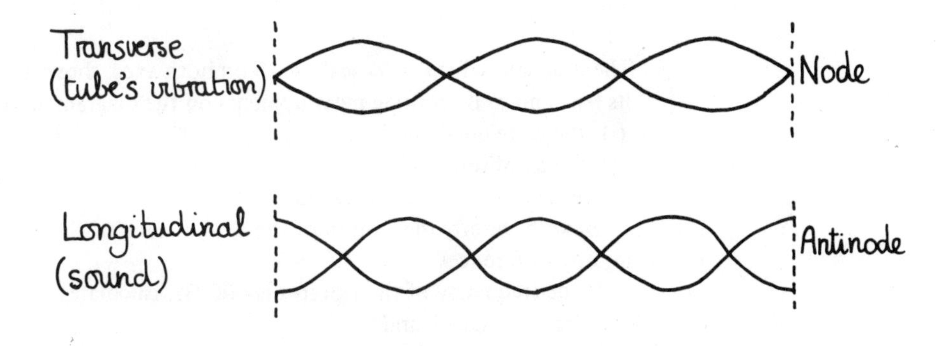

Example 3.8

Which one of the following must be heavily damped to function satisfactorily?

- **A** a simple pendulum
- **B** a sonometer wire
- **C** a tuning fork
- **D** a loudspeaker cone
- **E** a ballistic galvanometer (OLE)

Solution 3.8

A loudspeaker cone should vibrate at whatever frequency it is driven at by the signal. Any of its own natural oscillation should be damped away as quickly as possible. A loudspeaker that responded much better at one frequency would be useless.

Answer **D**

Example 3.9

In an experiment to investigate the properties of stationary waves, one end of a rubber cord is attached to a vibrator, the frequency of which can be varied, and the other end to a rigid support. The distance l, between the vibrator and the fixed support, can be varied.

(a)

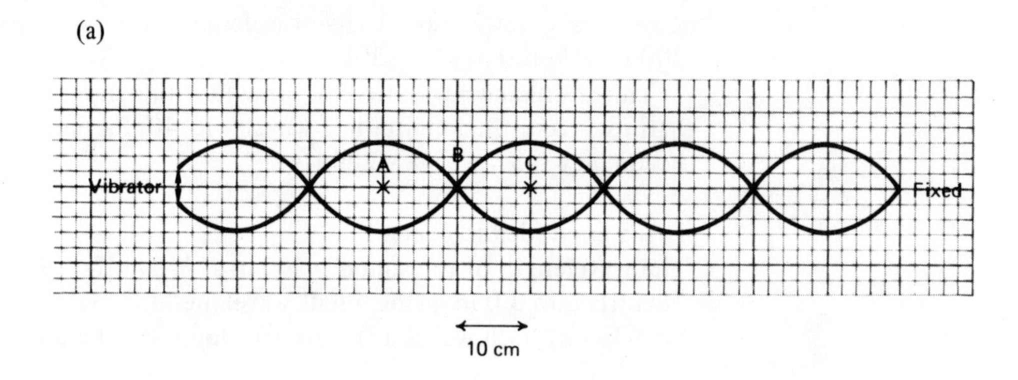

The diagram, which is to scale 1:10 on both axes, shows the cord vibrating at one of its harmonics. By making measurements on the diagram, determine

(i) the wavelength, and

(ii) the amplitude

of the stationary wave portrayed.

Describe briefly the motion of the cord at each of the points A, B, and C emphasising any differences. [4]

If the frequency of the vibrator is 400 Hz, calculate

(iii) the wave speed, and

(iv) the fundamental frequency of the cord when supported in this manner. [4]

(b) (i) A stationary wave system can be produced when two progressive waves interfere. Explain how this comes about in the cord when it is made to vibrate. Include diagrams to show the superposition of the waves involved to produce the resultant effect. [5]

(ii) The wave speed, c, in this experiment may be expressed as

$$c = \sqrt{\frac{Tl}{m}}$$

where T is the tension in the cord, l its length when under tension T and m is its mass. Show that the dimensions on the right hand side of the equation are those of speed. [2]

(iii) The unstretched length of the rubber cord is 0.8 m. When its length is extended to 1.0 m, the cord can be induced to vibrate in its fundamental mode and the wave speed is found to be 17.0 m s^{-1}. What will be the new wave speed if the cord is extended to 1.2 m? Assume that Hooke's law is obeyed by the cord. [3]

(L)

Solution 3.9

(a) (i) The node-to-node distance is $\lambda/2$, so the wavelength is 40 cm.

(ii) The zero-to-peak distance is the amplitude. Amplitude is 6 cm.

At A and C the string oscillates up and down. The points are in antiphase. As A is going up C is going down and vice versa. The string at B does not move. It is a node.

(iii) Wave speed = (frequency) × (wavelength)

$= (400 \text{ Hz}) (0.40 \text{ m}) = 160 \text{ m s}^{-1}$

(iv) At the fundamental frequency the entire string is half a wavelength.

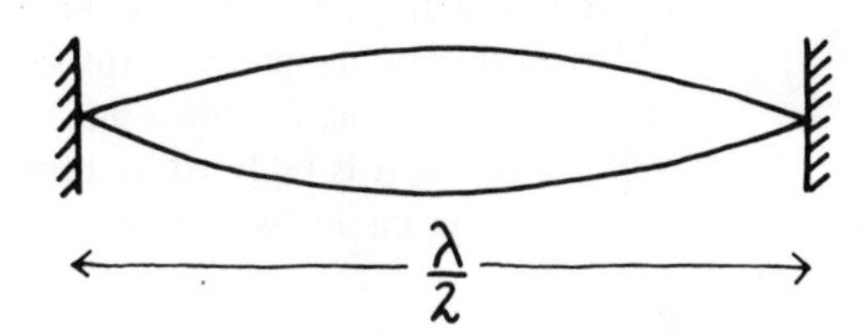

So wavelength = 2 (100 cm) = 200 cm = 2.0 m

$$\text{frequency} = \frac{\text{wave speed}}{\text{wavelength}} = \frac{160 \text{ m s}^{-1}}{2.0 \text{ m}} = 80 \text{ Hz}$$

(b) (i) The wave approaching the reflector interferes with the wave moving away from it. Because the waves are moving in opposite directions their phase relationship continuously varies. The diagrams below show them in phase and in antiphase.

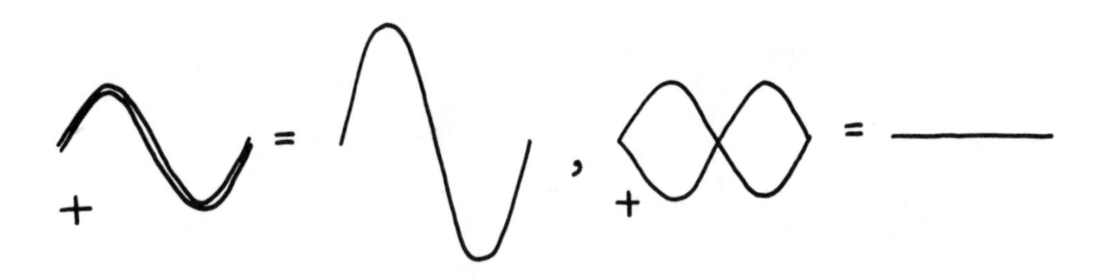

(ii) $\sqrt{\dfrac{Tl}{m}}$ has dimensions $\sqrt{\dfrac{(\text{MLT}^{-2})(\text{L})}{\text{M}}}$ and units $\sqrt{\dfrac{\text{N m}}{\text{kg}}}$

Newton's second law gives the units of force as kg m s^{-2}. Simplifying each, dimensions are $(\text{L}^{-2}\text{T}^{-2})^{\frac{1}{2}}$, units $(\text{m}^2 \text{ s}^{-2})^{\frac{1}{2}}$, i.e. the dimensions are LT^{-1} and the units m s^{-1}. These are clearly the dimensions and units of speed.

Give dimensions in the examination, but feel free to check equations using the units alone.

(iii) Tension when stretched from 0.8 m to 1.2 m is twice that when stretched to only 1.0 m.

The length is increased by a factor of 1.2 m/1.0 m = 1.2 and the mass is the same as before.

New wave speed = $(17.0 \text{ m s}^{-1}) (2 \times 1.2)^{\frac{1}{2}} = 26.3 \text{ m s}^{-1}$.

Example 3.10

(a) What conditions are necessary in order that interference patterns between light from two sources may be observed? [2]

(b) Draw a labelled diagram showing the apparatus required to determine the wavelength of red light using a pair of slits. Indicate approximate values for the dimensions of the apparatus and state a measuring instrument suitable for each measurement required. [7]

How would you use the measured values of the dimensions of the apparatus to estimate the separation of the fringes produced by light of wavelength 500 nm? [2]

What part is played by diffraction in this experiment? [2]

How are the fringes produced in this experiment very different from those produced using a diffraction grating and the same source of light?

(i) when the grating spacing is the same as the slit separation, and

(ii) when the grating spacing is much smaller than the slit separation?

If the red light is replaced by blue light, how do the fringes produced in the experiment differ markedly from those produced using red light? [7]

(L)

Solution 3.10

(a) Interference patterns using light can only be observed when two beams of light derived from the same source overlap. The two beams come from sources which then have a fixed phase relationship – they are coherent sources.

Coherence is much more easily understood if you think of two loudspeakers. Here the movements only remain in phase or remain exactly out of phase if the two speakers, the sources, are driven from the same signal generator. For light the way to produce two sources is to use a double slit.

(b)

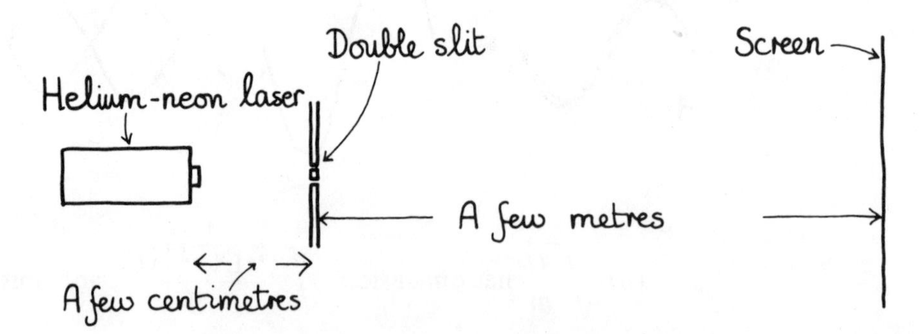

The double slit should have slits about 0.05 mm wide and about 0.2 mm apart. The distance between the maxima and minima of the interference pattern on the screen, x, can be measured with a ruler.

The double slit to screen distance, D, can be measured with a ruler.

The distance between the centres of the slits, a, can be measured with a travelling microscope.

For light of wavelength λ, $\lambda \approx ax/D$, so if $\lambda = 500$ nm

$$x = \frac{D}{a}\,(5.0 \times 10^{-7}\text{ m})$$

It is useful to give symbols for things when you mention them. For example x, D and a above. You can use them later as abbreviations.

The diffraction of light at each of the narrow slits produces a region where two coherent beams overlap.

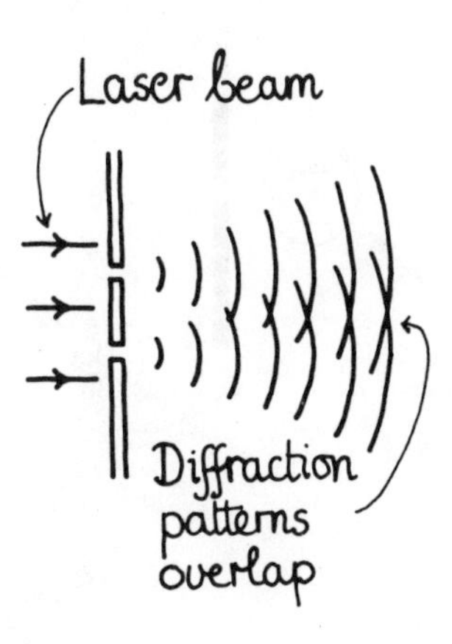

Do not draw with a compass – it takes far too long.

(i) The fringes produced by the diffraction grating coincide with the maxima of the fringes in the two-slit experiment, but they are very much brighter and are very narrow with darkness in between.

(ii) When the grating spacing is much smaller, the maxima of the fringes are much further apart, x is proportional to $1/a$ from the equation for x.

When blue light is used the maxima of the two-slit experiment are closest together, x is proportional to λ from the equation for x.

To answer questions describing fringes it is sometimes worth trying to sketch labelled graphs; but here they are not easy to do. You would need to draw

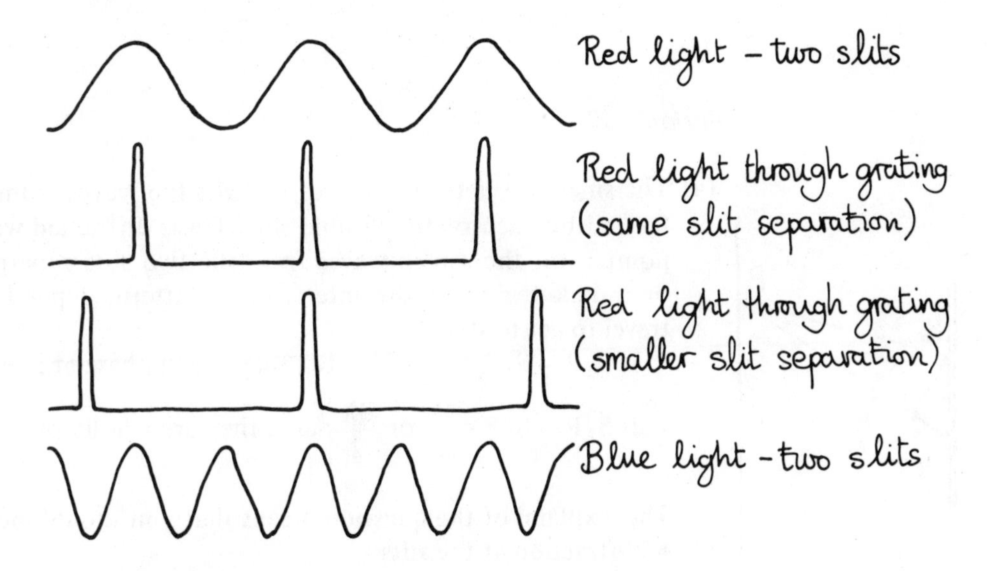

and add that the intensity of the grating graphs was much greater than the intensity of the two-slit graphs.

Example 3.11

The diagram (not drawn to scale) shows the apparatus used in an attempt to measure the wavelength of light using double slit interference.

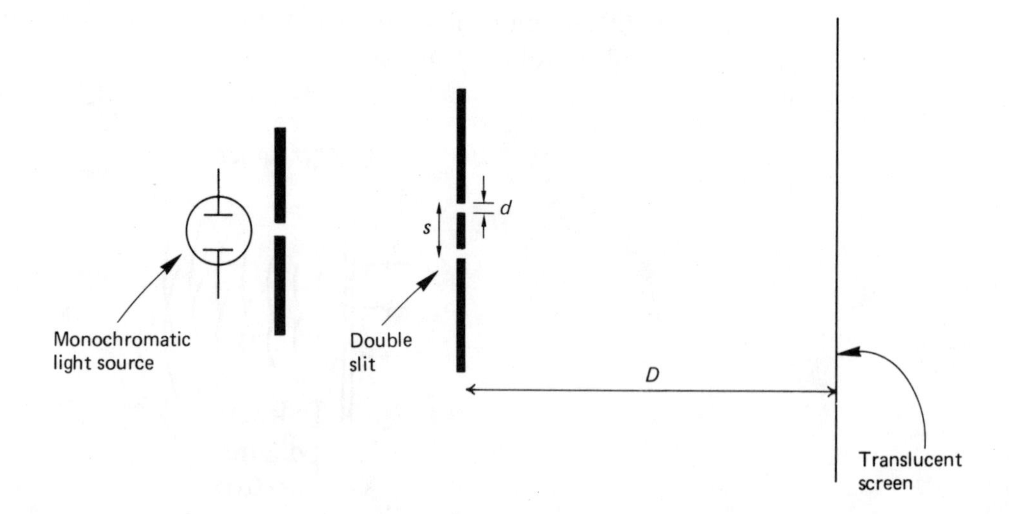

(a) Explain how the apparatus produces double slit interference fringes on the translucent screen. [4]

(b) If measurable fringes are to be seen on the screen, then
   (i) the slit separation s must be small;
   (ii) the distance D between the slits and the screen must be large.
   Explain each of these conditions. [4]

(c) The light from the monochromatic source has a wavelength of 5.0×10^{-7} m and it is required to produce a fringe separation of 5.0 mm. Suggest suitable values for s and D justifying your answer.
   Describe the important features of the pattern you would hope to obtain. [7]

(d) Describe and explain any changes in the pattern which would be brought about by each of the following changes made separately:
   (i) one slit is covered with an opaque material;
   (ii) the slits are made narrower although their separation remains the same. [6]

(AEB 1984)

Solution 3.11

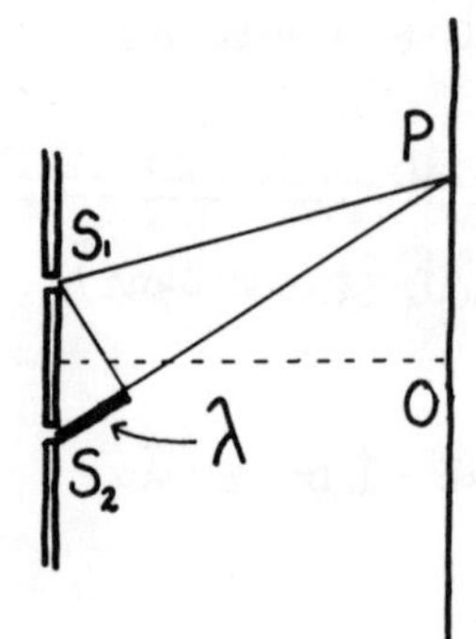

(a) The single slit acts as a source of light the waves from which are then diffracted by each of the double slits. These diffracted waves overlap and at a point P on the screen waves from the two slits superpose to give a bright or a dark region of the interference pattern, depending on how far they travel to get to P.

If $S_2P - S_1P = \lambda$ or 2λ, etc., they are in phase at P → bright.

If $S_2P - S_1P = \dfrac{\lambda}{2}$ or $\dfrac{3\lambda}{2}$, etc., they are wholly out of phase → dark.

The 'explain' of the question means that you should mention
- **diffraction at the slits**
- **superposition (interference) at the screen**
- **phase difference for bright/dark**
- **path difference $S_2P - S_1P$**

(b) Measurable fringes means that P must be a long way from O when $S_2P - S_1P = \lambda$

But wavelength of light $= \dfrac{s}{D}$ (fringe separation) i.e. $$\lambda = \frac{s}{D}(\text{OP})$$

$\therefore$ For large OP, $\text{OP} = \dfrac{D\lambda}{s}$

(i) s must be small, and
(ii) D must be large.

(c) From above $\dfrac{s}{D} = \dfrac{\lambda}{\text{OP}} = \dfrac{5.0 \times 10^{-7}\ \text{m}}{5.0 \times 10^{-3}\ \text{m}} = 10^{-4}$

As D can be a few metres, s must be a few tenths of a millimetre. Values of $D = 2.0$ m and $s = 0.2$ mm would work.

A graph of the pattern would be like this:

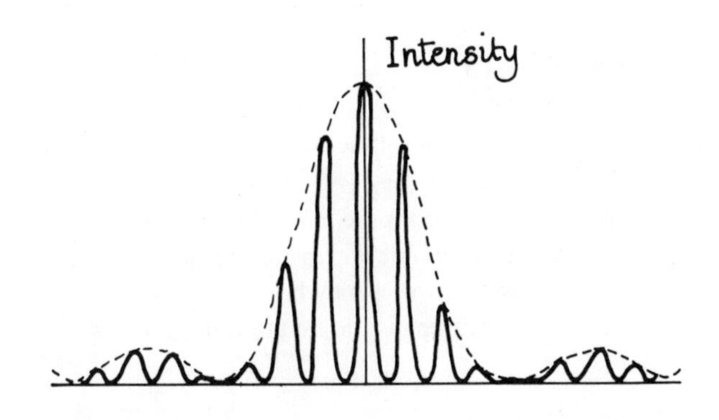

To describe this intensity pattern in words would involve mentioning the regularly spaced maxima, bright fringes, separated by minima, dark fringes, and the gradually decreasing intensity of the maxima to the sides of the centre of the pattern which is controlled by the single-slit diffraction pattern. The advantage of using the graph in this question is that changes in the pattern are then more easily described, and you can refer back to it and add to it as necessary.

(d) (i) The dotted line on the graph becomes the intensity pattern when one slit is covered. It is the diffraction pattern for a single slit of width d and light of wavelength 5×10^{-7} m.

(ii) If the slit separation remains the same, the maxima remain the same distance apart but the diffraction envelope spreads out sideways as each of the slits become narrower; $\sin \theta_m = \lambda/d$ gives the angle of the first minimum in this pattern for a slit of width d.

Example 3.12

A beam of light is shone straight at a diffraction grating with 2.5×10^5 lines per metre. The light from the grating is focused on a screen using a converging lens of focal length 40 cm. The wavelengths of two lines in the spectrum are 5.5×10^{-7} m and 6.0×10^{-7} m. What is the distance between the first-order images at these two wavelengths on the screen? [6]

Solution 3.12

The distance between two adjacent slits $= \dfrac{1}{2.5 \times 10^5\ \text{m}^{-1}}$

$$s = 4.0 \times 10^{-6}\ \text{m}$$

The angles at which the first-order spectra occur are found from $\lambda = s \sin \theta$

$$\therefore \sin \theta_1 = \frac{5.5 \times 10^{-7} \text{ m}}{4.0 \times 10^{-6} \text{ m}} = 0.138, \quad \text{and} \quad \sin \theta_2 = \frac{6.0 \times 10^{-7} \text{ m}}{4.0 \times 10^{-6} \text{ m}} = 0.150$$

These are small angles so we can approximate $\theta_1 = 0.14$ rad (about 8°) and $\theta_2 = 0.15$ rad.

A lens focuses parallel sets of rays in its focal plane at the centre of which is the principal focus. So the screen must be 40 cm from the lens. Thinking of the undeviated rays that go through the centre of the lens,
the image separation = (0.15 – 0.14) (0.40 m) = 0.004 m.

See Fig. 3.9 on page 48.

Example 3.13

Which one of the following statements concerning transverse waves is incorrect?

A Transverse waves may be polarised
B The oscillations are in a plane at right angles to the direction of propagation
C Electromagnetic waves are transverse waves
D Transverse waves may be set up in a vibrating wire
E Sound waves in a liquid are transverse waves (AEB 1984)

Solution 3.13

Electromagnetic waves are transverse because the magnetic and electric fields are at right angles to the direction of propagation. In sound waves the molecules move back and forth in the same direction as that in which the waves are travelling. So the last statement is incorrect.
Answer E

Example 3.14

(a) What are the approximate wavelength limits of the visible and X-ray regions of the electromagnetic spectrum? [2]

(b) The basic components of a spectrometer for investigating spectra in any region of the electromagnetic spectrum include a way of producing a parallel beam, a dispersive element and a detection system.

(i) The dispersive element in a spectrometer for use in the visible region is commonly a diffraction grating. Use the grating equation, $n\lambda = d \sin \theta$, to estimate a typical separation for the lines on such a grating. Could such a grating be used successfully in an X-ray spectrometer? Explain your answer. [3]

(ii) Name one detection system used in the visible and one used in the X-ray region of the spectrum. [2]

Solution 3.14

(a)

	Maximum	Minimum
Visible	7×10^{-7} m	4×10^{-7} m
X-rays	10^{-9} m	10^{-16} m

Doing a table is convenient and saves writing.

(b) (i) If the first-order spectrum is at $\theta = 20°$ and the light is of wavelength 5.5×10^{-7} m, then using

$$n\lambda = s \sin\theta, \qquad \text{where } n = 1$$

$$5.5 \times 10^{-7} \text{ m} = s \sin 20°$$

$$\Rightarrow \qquad s = 1.6 \times 10^{-6} \text{ m} \quad \text{or} \quad 1.6\ \mu\text{m}$$

Such a grating would just work with very long wavelength X-rays, but for $\lambda = 10^{-10}$ m the first-order spectrum would be at

$$\theta = \sin^{-1} \frac{1 \times 10^{-10} \text{ m}}{1.6 \times 10^{-6} \text{ m}} = 0.0036°$$

(ii) Photographic film can be used in both regions.

Example 3.15

A beam of electromagnetic waves of wavelength 3.0 cm is directed normally at a grid of metal rods, parallel to each other and arranged vertically about 2.0 cm apart. Behind the grid is a receiver to detect the waves. It is found that when the grid is in this position, the receiver detects a strong signal but that when the grid is rotated in a vertical plane through 90°, the detected signal falls to zero. What property of the wave gives rise to this effect? Account briefly in general terms for the effect described above. [4]

(L)

Solution 3.15

The property of polarisation gives rise to this effect.

All electromagnetic (e.m.) waves are transverse: the oscillations take place in directions at right angles to the direction in which the wave is travelling.

Two separate components of oscillation are possible: vertical and horizontal. E.m. waves consist of a changing electric field at right angles to a changing magnetic field. Both fields are at right angles to the direction of travel of the wave.

The direction of the electric field is what decides the polarisation. In the question the e.m. wave is horizontally polarised because when the rods are parallel to the changing electric field the signal is zero. This is because the rods act like receiving aerials and currents begin in the direction of the electric field. The changing currents in the rods then act like transmitters and the e.m. wave energy is reradiated in all directions. The power in the forward direction is therefore much less.

A dipole aerial actually receives and transmits in all directions except the direction in which the rod itself points. The principle is used in some radio direction finders.

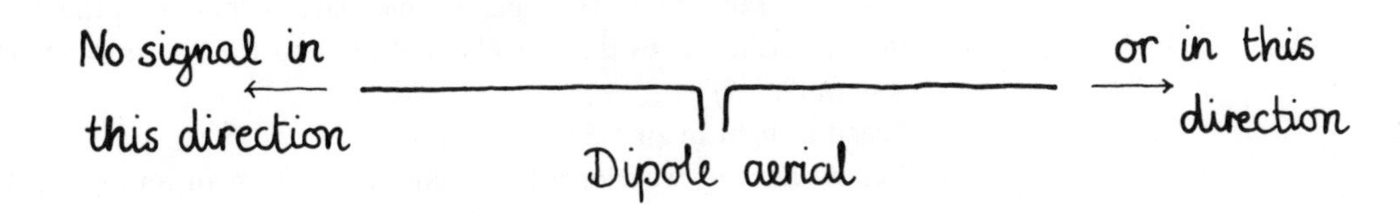

Example 3.16

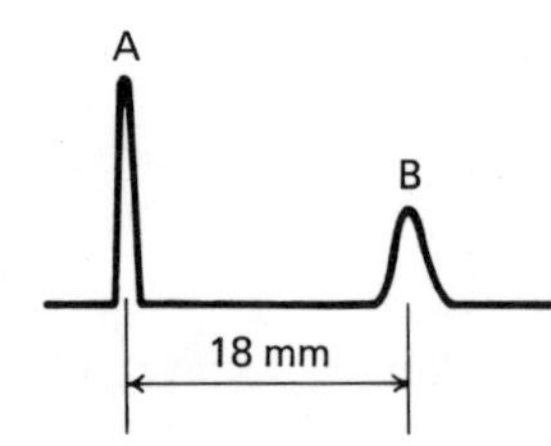

An oscilloscope is used to measure the time it takes to send a pulse of charge along a 200 m length of coaxial cable and back again. The diagram shows the appearance of the oscilloscope screen. A indicates the original pulse and B the same pulse after reflection.

If the time base speed is set at 10 mm μs^{-1}, calculate the speed of the pulse along the cable. [3]

(L)

Solution 3.16

Total distance = 2 x 200 m = 400 m

It is easy to forget the 2.

$$\text{Time taken for the pulse to return} = \frac{18 \text{ mm}}{10 \text{ mm } \mu s^{-1}} = 1.8\ \mu s$$

Use common sense or the units if you are not sure about the formula.

$$\text{Speed of pulse} = \frac{400 \text{ m}}{1.8 \times 10^{-6} \text{ s}} = 2.2 \times 10^{8} \text{ m s}^{-1}$$

One would expect a figure a little less than the speed of electromagnetic waves in a vacuum.

3.8 Questions

Question 3.1

Explain the meaning of (a) displacement, (b) acceleration. Define simple harmonic motion in terms of these quantities. [4]

A mass m is suspended by a light spring of constant k (the force required per unit extension). Show that the mass can perform vertical oscillations which are s.h.m.

Find expressions for the frequency, f, of oscillation (c) in terms of m and k, (d) in terms of g, the acceleration of free fall, and e, the extension of the spring when the mass hangs in equilibrium. [5]

Question 3.2

(a) The edge of a disc has 500 square teeth with alternate square slots. A parallel beam of light passes through one slot, strikes a mirror positioned 25 km away and returns along the same path. How many times each second does the disc rotate, if on returning, the light passes through the slot next to the one through which it passed on the way out?
Speed of light in air = 3.0×10^{8} m s^{-1}. [5]

(b) Yellow light of wavelength 560 nm passes from air into glass. If the glass's refractive index for yellow light is 1.5, calculate
(i) the speed of light in the glass
(ii) the wavelength of light in the glass
(iii) the frequency of light in the glass
(iv) the energy of the yellow photons in the glass. [5]
The Planck constant = 6.6×10^{-34} J s.

Question 3.3

(a) A and B are two sound sources emitting notes of the same frequency and amplitude. Describe and explain qualitatively what an observer would hear as he moves along the line XY.

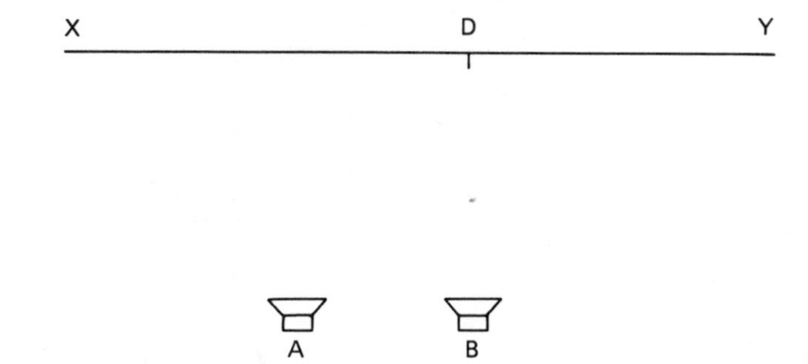

(b) The frequency of one of the sources is now changed slightly. Describe and explain qualitatively what an observer would hear if he remained stationary at some point D on XY. [9]

(AEB 1983)

Question 3.4

An amateur astronomer has a telescope of aperture 0.10 m and focal length 0.80 m.

(a) He is observing the full Moon which subtends an angle of 0.5° or 9×10^{-3} radian at the naked eye.

What is the diameter of the real image of the Moon in the focal plane of the telescope?

(b) He now tries to observe two stars close to each other which he has seen in a photograph taken with the 5 m aperture Mount Palomar telescope. But he only sees what appears to be a single star through his own telescope. Explain this observation.

Question 3.5

Monochromatic light, incident normally on a narrow slit S, is diffracted. A screen PQ is set up some distance from the slit in order to study the diffraction pattern.

(a) Sketch a graph of intensity I against distance x from the central point O along the line PQ on the screen.

(b) Describe qualitatively what happens to the diffraction pattern as the width of the slit is reduced. (Assume that it is practicable to reduce the slit-width until it is equal to a few wavelengths of the incident light.) [4]

Question 3.6

A parallel beam of red laser light of wavelength 600 nm falls normally on a single slit of width 0.10 mm. The pattern observed on a screen a few metres from the slit is represented graphically below. The vertical axis represents the intensity of the pattern. The horizontal axis indicates the position of the pattern on the screen.

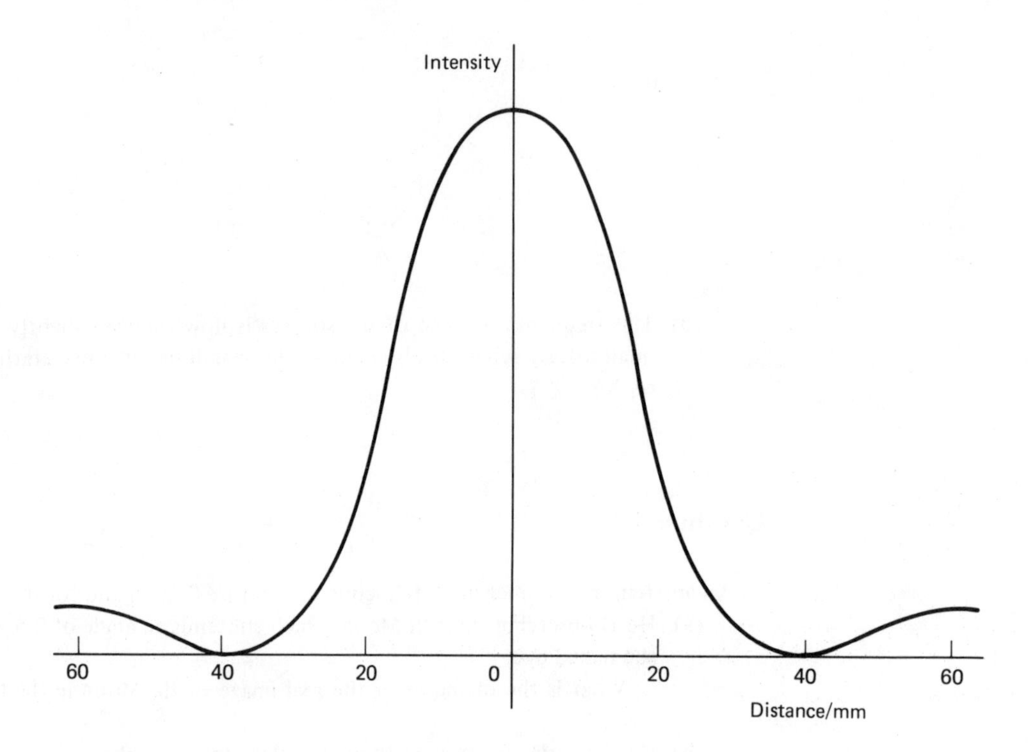

(a) Calculate the distance from the slit to the screen. [2]

(b) On axes of the same size draw the pattern that would be observed if the slit were halved in width. [3]

(c) The single slit is replaced with a double slit. The slits are both 0.10 mm wide and parallel. The distance between their centres is 0.40 mm. On a new set of axes with the x-axis the same as before, draw the graph representing the intensity pattern that would be observed. Show your calculations and indicate any change you have made in the intensity axis scale. [7]

Question 3.7

You have been supplied with a diffraction grating and you know it has between 400 and 4000 lines per cm. Describe how you would discover the figure more exactly if you were supplied with a sodium lamp producing yellow light of known wavelength, a large piece of card with a millimetre wide slit in it and a metre rule. The description of the experiment you would perform should include a diagram, an explanation of how you would calculate the result and a simple indication of the uncertainty in the final result. [10]

Question 3.8

An experiment is set up so that microwaves are reflected from a metal plate. The incident waves and the reflected waves interfere.

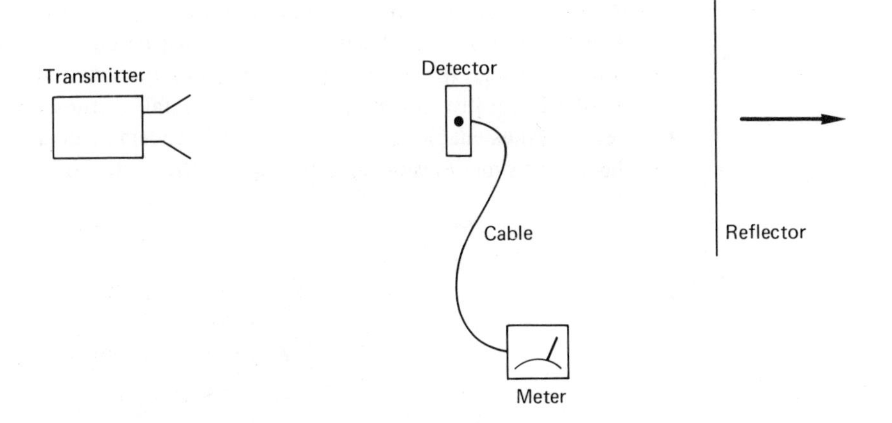

The reflector is moved slowly away from the detector and the reading on the meter is noted.

A maximum reading is found with the reflector at 22.5 cm from the detector and a further ten maxima are found as the reflector is moved to a maximum reading at 36.5 cm from the detector.

(a) Calculate the wavelength of the microwaves.

(b) What is the frequency of these microwaves? [4]

(SEB)

Question 3.9

This passage is adapted from an article by A. H. Cook in the *Proceedings of the Royal Institution*, volume 43. Read the passage and then answer the questions at the end.

The times taken by elastic waves from an earthquake to travel through the Earth give much information about its constitution. A very simple observation similar to one in optics shows us that there is a concentration of mass at the centre of the Earth.

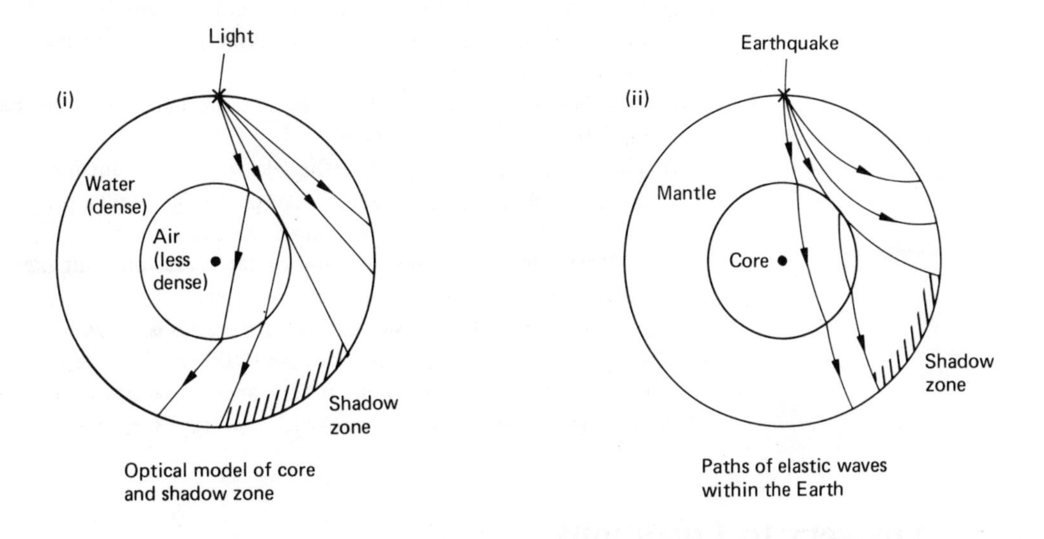

Optical model of core and shadow zone

Paths of elastic waves within the Earth

Figure (i) shows an optical model consisting of a cylinder filled with water within which there is an optically less dense cylinder. With a light at the outer edge of the water cylinder there is a shadow zone where no light can reach the edge. This is because of the way the less dense cylinder refracts the light. A similar thing happens in the Earth. There is a region between about 100° and 140° from an earthquake (measured around the surface of the Earth) within which no direct earthquake waves are received – see figure (ii).

From this observation it is possible to work out that there is a central core which is just over half the radius of the Earth. This core must be liquid because no transverse waves are transmitted through it and because the speed of the longitudinal waves within it is less than in the solid Earth just outside the core. More detailed studies of the times of travel of waves from earthquakes enable us to work out in considerable detail the way in which the density and the Young's modulus of the solid parts of the Earth vary.

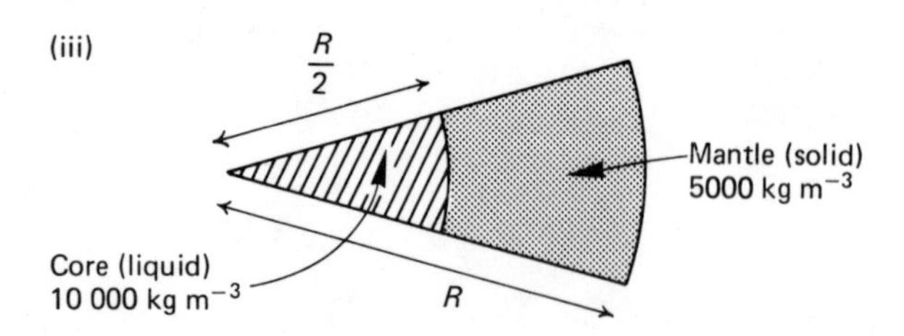

Figure (iii) shows very roughly how the density of the Earth varies, the solid mantle having half the density of the liquid core. The average density of the Earth, deduced from measurements of G, is 5600 kg m^{-3}, which is consistent with this data.

Questions

(a) (i) Both transverse and longitudinal waves are mentioned. Explain the difference between them.

(ii) Why can transverse waves not be transmitted through the central core of the Earth? [3]

(b) (i) There is a mistake in the passage and in figure (i). The optical model should have a central region in which there is an optically more, not less, dense cylinder, for example, glass.

Explain why this is necessary to explain the production of the shadow zone in figure (i).

(ii) Explain in your own words why the (corrected) arrangement of figure (i) is referred to as an optical model for earthquake waves. In what way is it an imperfect model?

(iii) Estimate the refractive index of the central region to produce the optical effect shown in figure (i) given that the refractive index of water is 1.3. Explain how you made your estimate. [8]

(c) (i) Does figure (ii) support the view that the density of the Earth's mantle is not uniform? Explain your answer.

(ii) The speeds of longitudinal earthquake waves just outside and just inside the core are 13.5 km s^{-1} and 8.0 km s^{-1} respectively. Draw a sketch showing wavefronts meeting the core-mantle boundary at an angle of 45°.

Mark the relative wavelengths of the waves on your sketch and calculate the angle at which the waves leave the boundary. [7]

(d) A simple view of how the density of the Earth varies is shown in figure (iii). How can the numbers given on this diagram (density core 10 000 kg m^{-3}, density mantle 5000 kg m^{-3}) be consistent with the known value for the average density of the Earth given at the end of the passage? Illustrate your answer with a (rough) specimen calculation. [3]

3.9 Answers to Questions

3.1 (a) and (b) Your explanation should be a definition or a defining equation with a comment about the vector/scalar properties for each.

(c) See worked example 3.4 for a proof of $T = 2\pi\sqrt{\dfrac{m}{k}}$ in this situation.

(d) Draw a diagram of the equilibrium position and show that $\frac{m}{k} = \frac{e}{g}$. Note that you are asked for the frequency of the oscillations in both cases.

The key to the problem is that the maximum velocity of a body undergoing s.h.m. is ωx_0 or $2\pi f x_0$ when the amplitude is x_0.

3.2 (a) 12.

(b) (i) 2.0×10^8 m s^{-1}.

(ii) 370 nm.

(iii) The frequency in and out of the glass must be the same otherwise a backlog of oscillations would build up! So the answer is 5.4×10^{14} Hz.

(iv) The photons do not lose energy when passing into the glass. They would need to gain it again on the way out to avoid a change of colour. The answer is 3.5×10^{-19} J

3.3 (a) This is the basic two-source superposition experiment producing an interference pattern. A good answer will have sketches explaining constructive and destructive interference.

(b) This is a beat phenomenon. Call the two frequencies f_1 and f_2 in your answer. A graph of the intensity of the sound heard during beats is like a displacement–time graph for a coupled oscillator, the shape of the intensity graph resulting from superposition between the two waves. See worked example 3.6.

3.4 (a) The focal length is what affects the magnification.
Answer: 7 mm

(b) The aperture affects the resolution and brightness of the image.

All stars except the Sun are effectively point sources when seen from Earth. Their apparent diameter is due to diffraction at the telescope aperture.

3.5 (a) Be careful that you draw a curve for a single-slit pattern of diffraction and not the interference pattern for a two-slit arrangement which has a diffraction envelope. See Question 3.6.

(b) Though a qualitative answer is asked for you should have $\sin\theta = \lambda/d$ for the first minimum in mind.

3.6 (a) 6.7 m.

(b) Remember the intensity will decrease and that the pattern will get wider.

(c) Draw the single-slit diffraction pattern faintly and then draw the interference pattern within this envelope. Calculate the positions of the interference maxima with the formula that looks the same as the formula you have already used for a single-slit diffraction minimum. See worked example 3.11.

3.7 The slit can be illuminated from behind and observed from about a metre away through the grating close to the eye. One will see several images of the slit at various angles. They appear to be either side of the slit itself.

3.8 (a) Remember the distance between antinodes is half a wavelength. So the wavelength is 2.8 cm.

(b) Microwaves are electromagnetic waves. The frequency is 11 GHz.

3.9 In answering questions like this one, you should not have spent too much time reading and re-reading the passage.

(b) (i) Light rays bend towards the normal when passing from an optically less dense to an optically more dense medium, that is from low to high refractive index.

(ii) Models help us to explain behaviour and, perhaps, to make predictions.

(iii) You will have to measure angles from the diagram and use Snell's law in the form $n_w \sin \theta_w = n_g \sin \theta_g$. Choosing θ_w to be 90° makes it easier.

Answer around 1.4–1.5

(c) (ii) The ratio of the wavelengths should be the ratio of the wave speeds. The angle the wavefronts make with the core–mantle boundary after refraction is 28°.

(d) You need to use $\frac{4}{3}\pi r^3 \rho$ for the mass of a sphere.

4 Electricity and Electronics

4.1 You Should Recognise

Quantity	*Symbol*	*Unit*	*Comments*
electric current	I	ampere, A	
electric charge	Q	coulomb, C	1 C = 1 A s
electronic charge	e	coulomb, C	a constant = 1.6×10^{-19} C
current density	J	ampere per metre squared, A m^{-2}	
electrical energy	W	joule, J	1 kW h = 3.6×10^6 J
electrical potential difference	V	volt, V	often abbreviated to p.d.; 1 V = 1 J C^{-1}
e.m.f.	E	volt, V	
electrical power	P	watt, W	1 W = 1 J s^{-1}
resistance	R	ohm, Ω	often kΩ or MΩ
internal resistance	r	ohm, Ω	for cells often negligible
resistivity	ρ	ohm metre, Ω m	*not* Ω/m; for copper 2×10^{-8} Ω m
conductivity	σ	siemens, S	1 S = Ω^{-1} m^{-1}
capacitance	C	farad, F	often μF or pF
time constant	τ	second, s	equal to RC
relative permittivity	ϵ_r	(no unit)	usually in the range 2-6; 1.0006 for air
permittivity of vacuum	ϵ_0	farad per metre, F m^{-1}	a constant = 8.85×10^{-12} F m^{-1}

Not listed are quantities, such as mass, length and time, which occur in all chapters and which are given in the list on page 7.

4.2 You Should be Able to Use

- Electric current:
  Kirchhoff's first law $I = I_1 + I_2$ at a branched electric circuit
  current as a rate of flow of charge $I = \dfrac{Q}{t}$ or $\dfrac{dQ}{dt}$
  drift speed, v, of electrons in a wire $I = nAve$ or $J = nve$
  where n is the number of electrons per cubic metre

- Electrical energy:

  p.d. is either $\dfrac{\text{energy transfer}}{\text{charge}}$ or $\dfrac{\text{power converted}}{\text{current}}$

  $$V = \frac{W}{Q} \quad \text{or} \quad V = \frac{P}{I} \quad \text{i.e. } P = IV \text{ (watt = amp × volt)}$$

  e.m.f. is defined in the same way but see section 4.4(c)

- Resistors and resistance:

  resistance $R = \dfrac{V}{I}$, $V = IR$

  Ohm's law $V \propto I$ (or R = constant) for metals at constant temperature

  power in resistors $P = I^2R = \dfrac{V^2}{R}$

  resistance and resistivity $R = \rho \dfrac{l}{A}$

  temperature coefficient of resistance, α, for metals $R = R_0\ (1 + \alpha\Delta\theta)$

- Circuit calculations:

  resistors in series $R = R_1 + R_2 + R_3 + \ldots$

  resistors in parallel $\dfrac{1}{R} = \dfrac{1}{R_1} + \dfrac{1}{R_2} + \dfrac{1}{R_3} + \ldots$

  cells in series total e.m.f. $= E = E_1 + E_2 + \ldots$

  cells of the same e.m.f. in parallel total e.m.f. $= E$

  for a simple circuit total e.m.f. = (current) (total resistance)

  $E = I\,(R + r)$

  Kirchhoff's second law $\Sigma E = \Sigma IR$ round any closed circuit loop

- Capacitors and capacitance:

  capacitance $C = \dfrac{Q}{V}$, $Q = CV$

  energy stored in a capacitor W = (average p.d.) (charge)

  $= \frac{1}{2}QV = \frac{1}{2}CV^2$

  capacitors in series $\dfrac{1}{C} = \dfrac{1}{C_1} + \dfrac{1}{C_2} + \dfrac{1}{C_3} + \ldots$

  capacitors in parallel $C = C_1 + C_2 + C_3 + \ldots$

  parallel-plate capacitors $C = \epsilon_0 \dfrac{A}{d}$ in air

  $C = \epsilon_0 \epsilon_r \dfrac{A}{d}$ with dielectric

- Charging and discharging capacitors through resistors in a simple R–C circuit:

  discharging $Q = Q_0 e^{-t/RC}$ initial charge Q_0

  charging $Q = Q_0(1 - e^{-t/RC})$ final charge Q_0

- Electronics:

  words representing logic systems NOT, OR, NOR, AND, NAND

4.3 Circuit Symbols

You will know many circuit symbols, such as those for a cell or a resistor, which are very common. Some less well known but important symbols include:

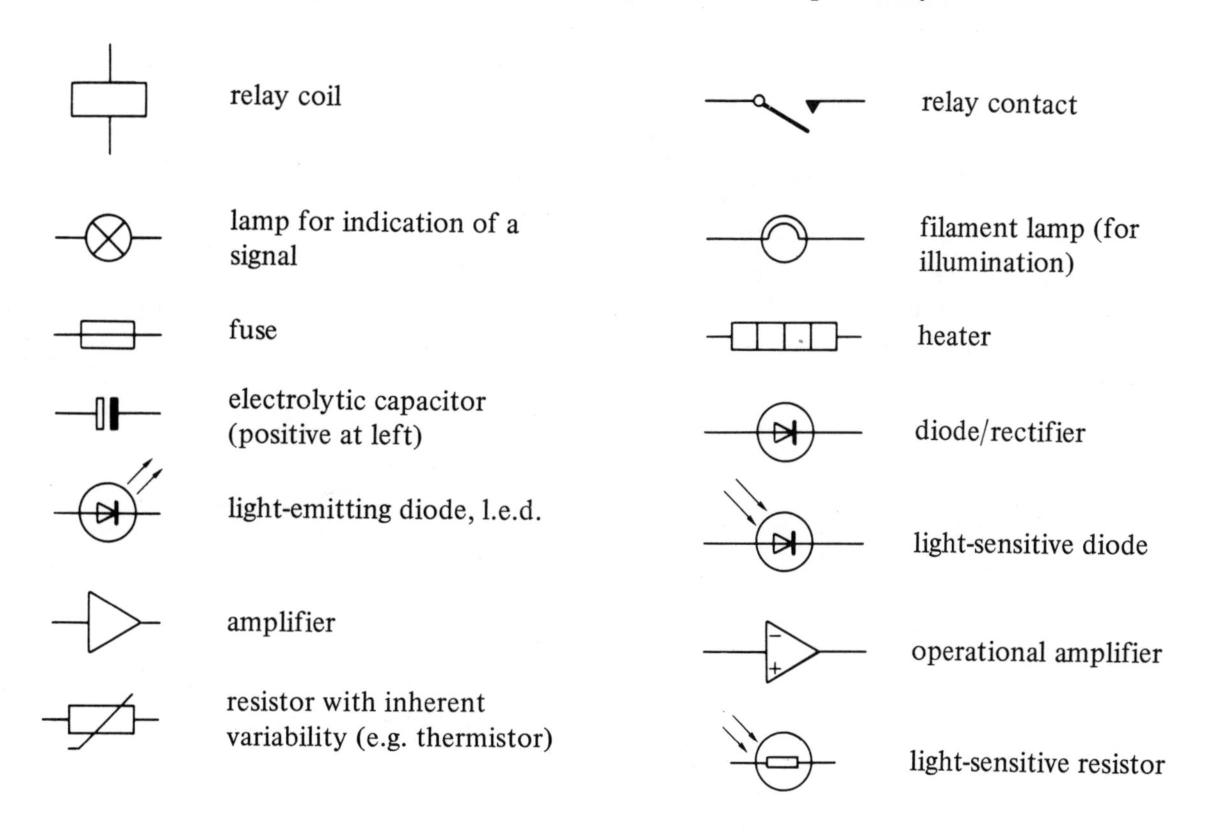

4.4 Properties of Electrical Components

(a) *I–V* Characteristics

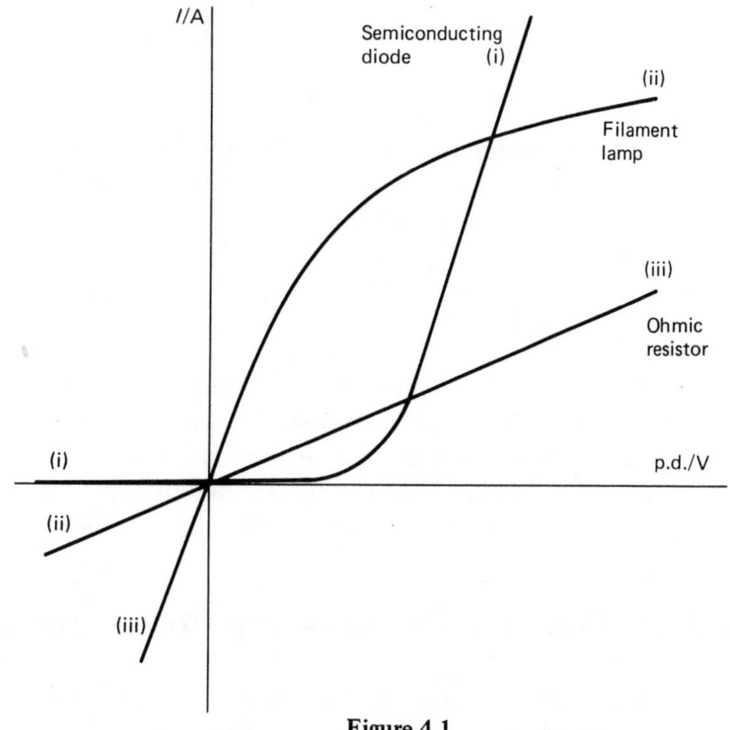

Figure 4.1

Characteristic curves for two non-ohmic conductors and one that obeys Ohm's law are shown in Fig. 4.1.

The resistance of all non-ohmic conductors varies; R for the diode *falls*, R for the lamp *rises*, as V increases.

(b) Resistors in Series and in Parallel

For two conductors in series, the current I is the same in each and $V = V_1 + V_2$ – see Fig. 4.2(a). The p.d.'s split in the ratio of the resistances for the particular current: $V_1/V_2 = R/R_t$.

For two conductors in parallel, the p.d. V is the same across each and $I = I_1 + I_2$ – see Fig. 4.2(b). The current splits in the inverse ratio of the resistances for the particular p.d.: $I_1/I_2 = R_d/R$.

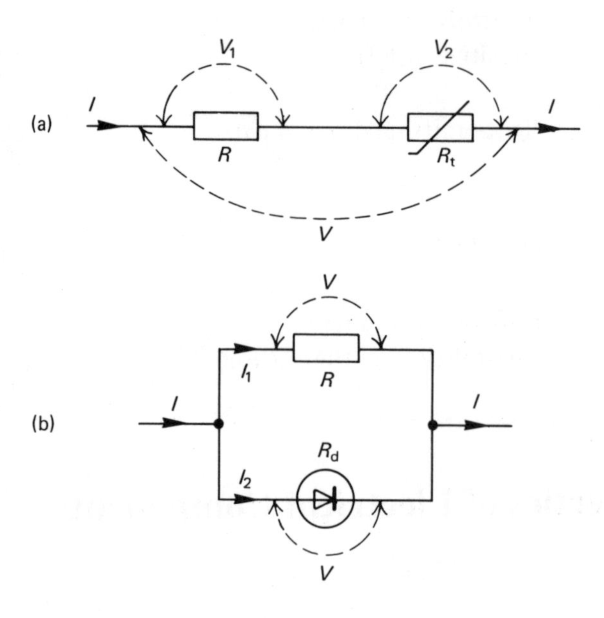

Figure 4.2

When considering power conversion for two conductors in series use I^2R, as the current is the same in each, but use V^2/R for two conductors in parallel, as then the p.d. is the same for each.

(c) p.d. and e.m.f.

The only difference between p.d. and e.m.f. is the direction of the energy conversion. In cells energy is converted from chemical energy into electrical energy. In all electrical components, including cells, energy is converted from electrical energy to other forms (often internal energy). As within a cell both processes are happening, the terminal p.d. is less than the e.m.f.: $E = V + IR$ (see Fig. 4.3).

4.5 Rheostats, Potential Dividers and Potentiometers

Ammeters have a small R and voltmeters a large R. Unless you are told otherwise, assume the values are zero and infinity respectively. Three commonly met circuits

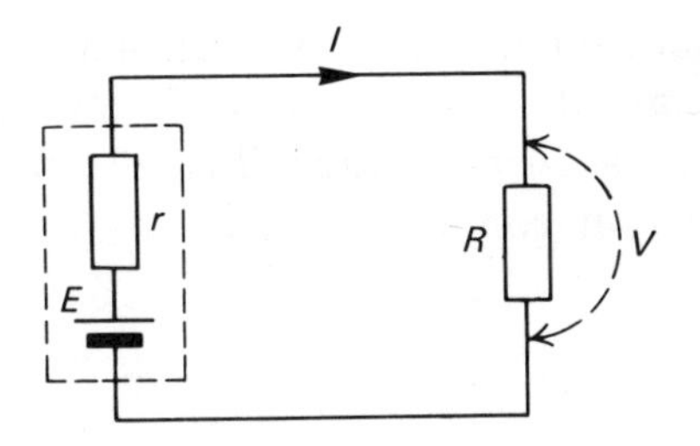

Figure 4.3

are shown in Fig. 4.4. In circuit (a) the minimum current in R is not zero but circuit (b) provides a way of varying the current in R all the way down to zero. In circuit (b) the p.d. across R for a given setting of the potentiometer can be found without calculating currents – see worked example 4.8.

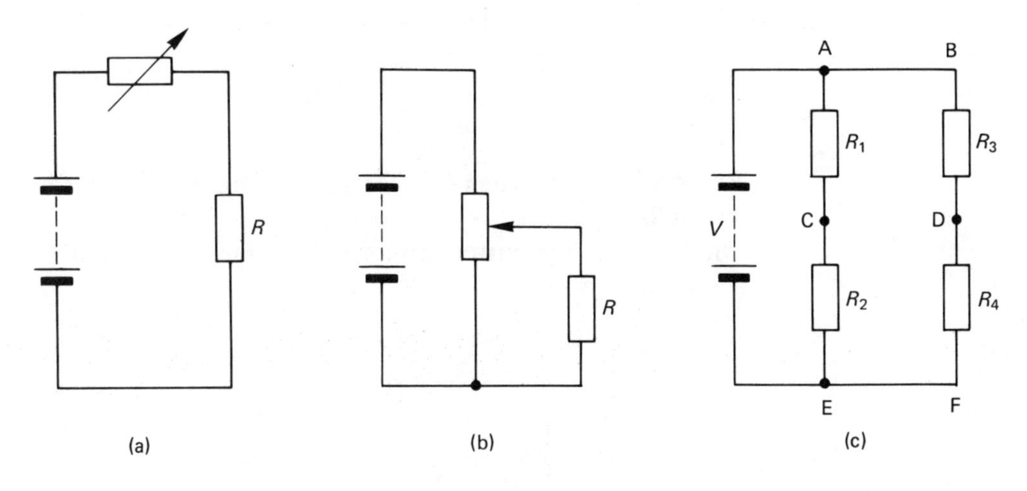

Figure 4.4

Circuit (c) should be 'seen' in terms of potentials and potential differences. Suppose we think of V_E and V_F at zero potential, e.g. that they are earthed. Then as

$$\frac{V_{CE}}{V_{AE}} = \frac{R_2}{R_1 + R_2}$$

you can write

$$V_C = \frac{R_2}{R_1 + R_2} V$$

A similar calculation for V_D then tells you the relative potentials at C and D. Circuits involving balanced potentials can be used to compare p.d.'s or e.m.f.'s.

4.6 Capacitors

A capacitor stores charge, $Q = CV$. If the p.d. across it rises by ΔV then the extra charge stored is $\Delta Q = C\Delta V$, and if this happened in a time Δt, then

$$\frac{\Delta Q}{\Delta t} = \frac{C\Delta V}{\Delta t} \qquad \text{i.e.} \qquad I = C\frac{\mathrm{d}V}{\mathrm{d}t}$$

Parallel-plate capacitors can have a variable capacitance. The charge, and hence energy, they store can be altered. If the energy rises then you can always find an external energy source; if it falls the energy can always be traced, often to internal energy in connecting wires.

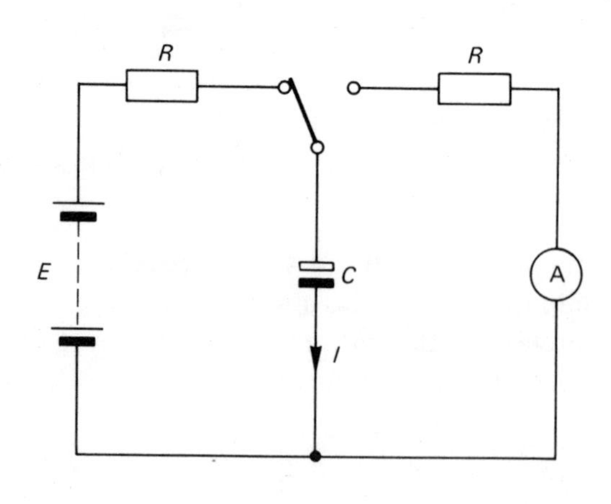

Figure 4.5

Refer to Fig. 4.5. When C is charged the initial current is high and equal to E/R; it then falls exponentially to zero.

On discharge the current follows the same pattern but in the opposite sense in C – see Fig. 4.6(a).

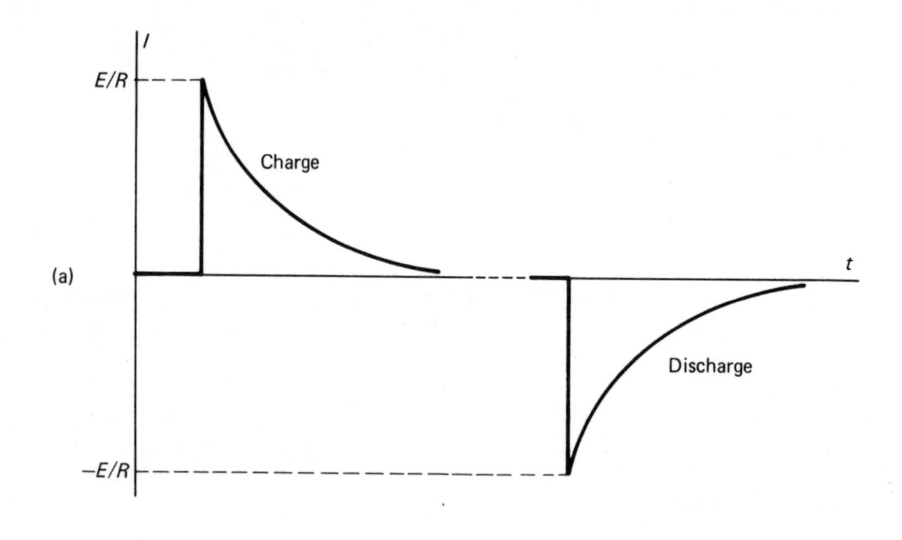

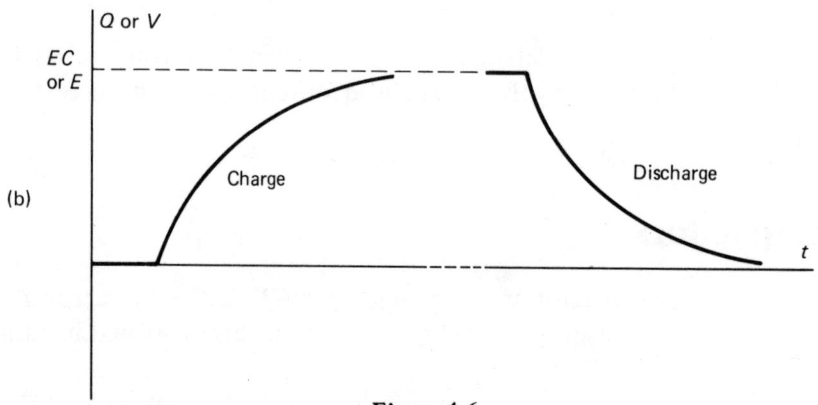

Figure 4.6

For both cases $I = I_0 e^{-t/RC}$.

After a time RC the current has fallen to $1/e$ of its initial value, after $2RC$ to $1/e^2$ etc.

The charge or p.d. across C vary as in Fig. 4.6(b). The formulas are given in section 4.2.

When you charge and discharge continuously, e.g. using a vibrating reed switch, the average discharge current can be measured and hence C found from $I_{av} = fCE$, a relationship you should be able to prove.

4.7 Electronics

(a) Analogue Circuits

(i) *Definition*

Physical quantities are represented by a potential difference that can have any value between an upper and a lower limit. In the case of a typical operational amplifier these limits are −15 V and +15 V. See Fig. 4.7.

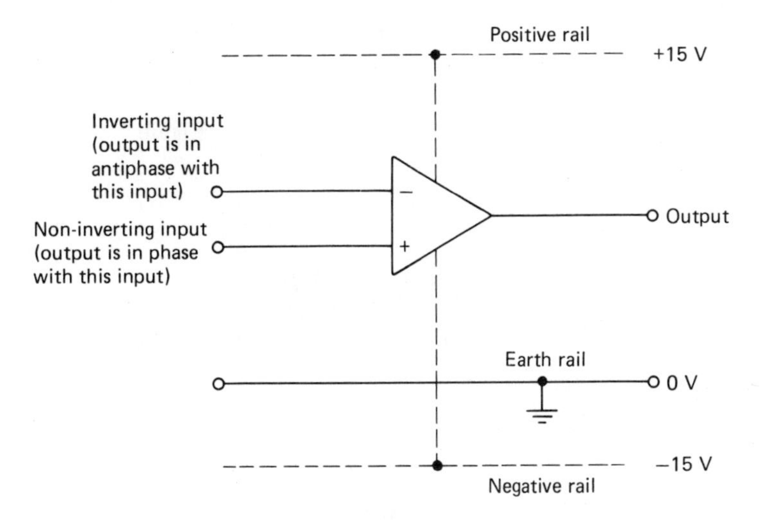

Figure 4.7

(ii) *The Operational Amplifier*

An operational amplifier is a multistage amplifier. Its 'silicon chip' construction makes it cheap, reliable, adaptable and above all compact. A good one has a very high inherent gain and a very high input resistance (properly called input impedance).

The dotted supply rails are often omitted from the circuit diagrams.

(iii) *Negative Feedback*

Negative feedback is an important concept in the study of all kinds of systems. In amplifiers it is achieved by feeding back a proportion β of the output signal into the input in such a way that it is in antiphase with the original input signal.

For an amplifier with a high inherent voltage gain A (e.g. 100 000)

$$A\,(V_i - \beta V_o) = V_o$$

V_o is the output voltage, V_i is the input voltage to the whole system and the quantity in brackets is the input to the op. amp. itself. Rearranging the formula it becomes:

$$\text{Gain of system} = \frac{V_o}{V_i} = \frac{A}{1 + A\beta} \approx \frac{1}{\beta} \qquad \text{since } A\beta \gg 1$$

A practical arrangement for feedback is shown in Fig. 4.8. The current through the two resistors is the same because there is almost no current into the op. amp. itself. Owing to the huge voltage gain of the op. amp. the potential of its inputs is almost zero, i.e. virtually earth potential.

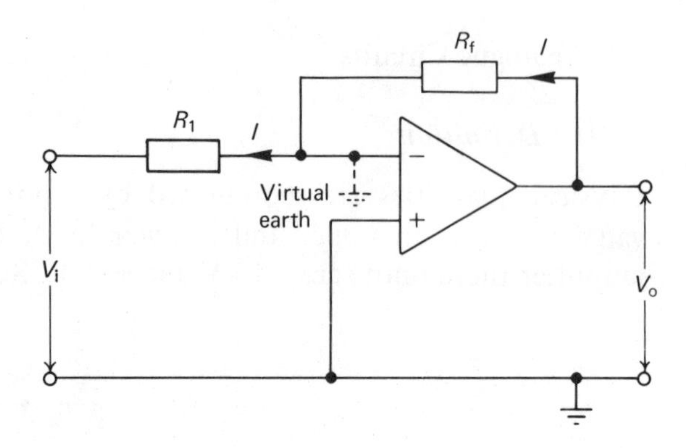

Figure 4.8

Using Ohm's law

$$V_i = -R_1 I \qquad \text{and} \qquad V_o = R_f I$$

The – sign occurs because of the direction of the current and the potential difference.

$$\text{Gain of the system} = \frac{V_o}{V_i} = \frac{-R_f}{R_1}$$

See the examples for other feedback circuits.

Negative feedback has the advantage of stabilising gain and making it constant over a wide band of frequency. Its disadvantage is that it reduces gain.

(iv) *Positive Feedback*

This occurs when a proportion of the output signal is fed back in phase with the input. A simple example is when the microphone of an amplification system is put too close to the loudspeaker and a wailing noise results. This basic effect can be improved on to produce oscillators with full control over the frequency and amplitude of their output.

(b) Digital Circuits

(i) *Definition*

Physical quantities are represented by numbers, which are in a binary code made up with a combination of different potentials that are either above a certain value (high or logic 1) or below that value (low or logic 0).

(ii) ***Logic Gates***

The truth table for several two-input gates is shown in Fig. 4.9. Notice that NOR is the inverse of OR. It means Not OR. More than two inputs are common in practice.

Input A	Input B	Output of OR	Output of NOR	Output of AND	Output of NAND	Output of EOR
0	0	0	1	0	1	0
1	0	1	0	0	1	1
0	1	1	0	0	1	1
1	1	1	0	1	0	0

Figure 4.9

The system for drawing gates is not yet standardised (Fig. 4.10) so if in doubt you should label them clearly. Notice the inverting gates have a little circle.

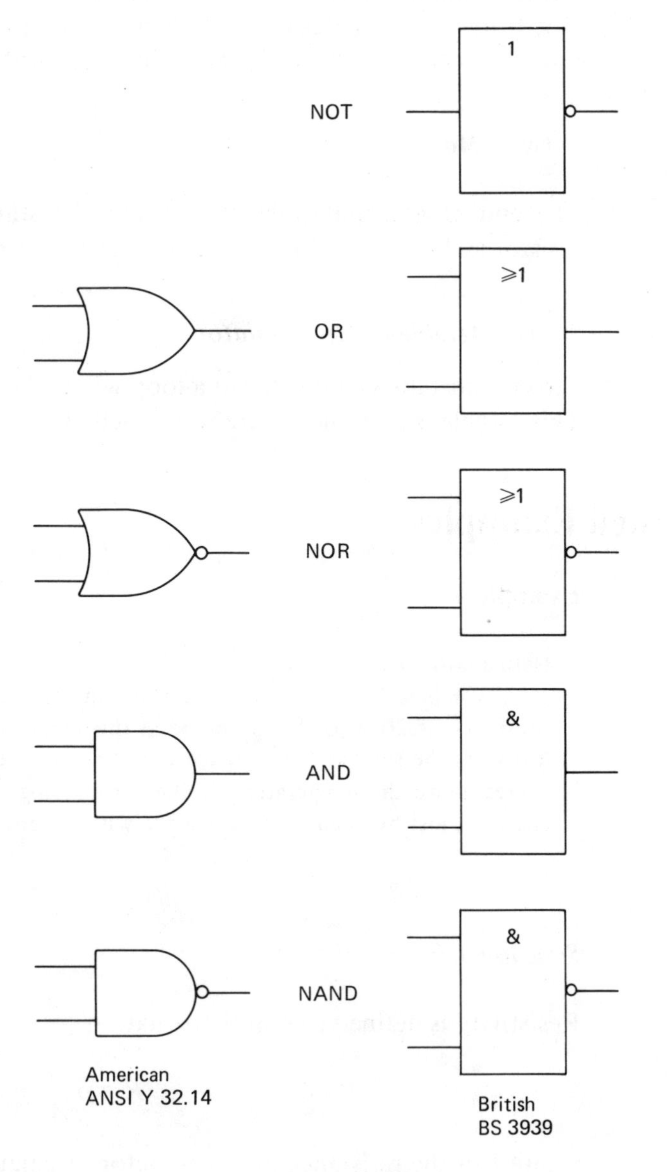

Figure 4.10

(iii) ***Bistable or Flip Flop***

It has two stable states and can be changed from one to the other by changing to the logic level of one of the set or reset connections (Fig. 4.11). Two NAND gates will also make a bistable, the difference being the steady logic level of the set and reset connections.

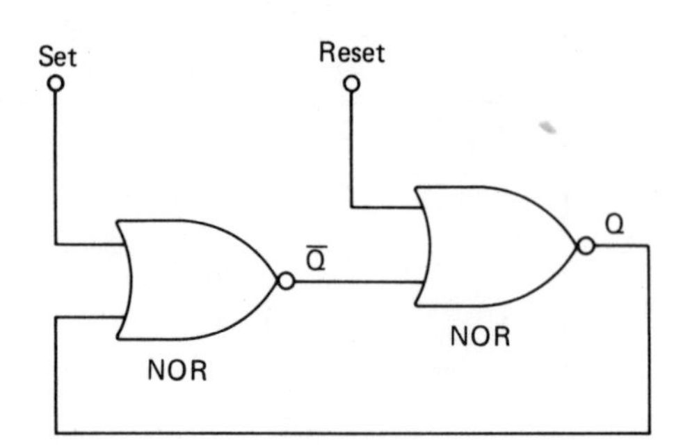

Figure 4.11

The bistable is a one-bit memory store.

A bistable can be designed to change state only at the END of an input pulse. A series of such bistables will make a binary counter.

(iv) ***Monostable***

The end of an input pulse will change its state, but after a time decided by a resistor and capacitor in the circuit it will change back. The time is about RC.

(v) ***Astable or Multivibrator***

Two monostables connected in a loop will make a circuit that continuously pulses. This astable is a simple square-wave oscillator.

4.8 Worked Examples

Example 4.1

Define *resistivity*.

A 3-ampere fuse is to be made from an alloy which melts at 230°C and which has a resistivity of 20×10^{-8} ohm metre at this temperature. It may be assumed that energy is lost from the surface of the fuse wire at 8.0 watts per square metre of surface area for every degree above the temperature of the surroundings which is 20°C. Assuming that no heat energy is lost by conduction along the wire, determine the diameter of fuse wire required.

[7]

(AEB 1983)

Solution 4.1

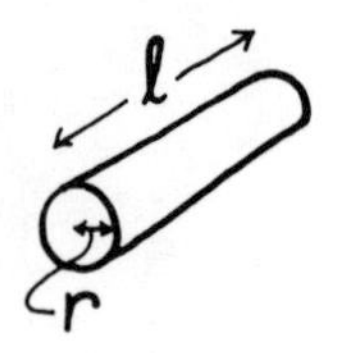

Resistivity is defined as ρ in the equation

$$R = \rho \frac{l}{A}$$

where R is the resistance of a conductor of length l and cross-sectional area A.

Use equations to define physical quantities

Fuse melts at 230°C, that is (230–20) K = 210 K above the surroundings. Under these conditions

power loss from the surface of the fuse = $(8.0 \text{ W m}^{-2} \text{ K}^{-1})(210 \text{ K})$

$$= 1680 \text{ W m}^{-2} = \frac{P}{A}$$

This is the difficult part of the question – interpreting what is meant by 8.0 watts per square metre of surface for every degree above the temperature of the surroundings.

If the fuse wire is of length l and radius r:

its resistance = $\rho \dfrac{l}{\pi r^2}$ and its surface area (curved) = $2\pi rl$

∴ Power converted in fuse for a current I is

$$P = I^2 R = I^2 \frac{\rho l}{\pi r^2}$$

and the power loss from its surface is

$$\frac{P}{A} = \frac{I^2 \rho l}{\pi r^2 \times 2\pi rl} = \frac{I^2 \rho}{2\pi^2 r^3}$$

As all the power converted is lost from the surface of the fuse

$$1680 \frac{\text{W}}{\text{m}^2} = \frac{I^2 \rho}{2\pi^2 r^3}$$

$$r^3 = \frac{(3 \text{ A})^2 (20 \times 10^{-8} \ \Omega \text{ m})}{2\pi^2 (1680 \text{ W m}^{-2})}$$

i.e. $r = 3.8 \times 10^{-4}$ m or 0.38 mm so the diameter of the wire is 0.76 mm.

Example 4.2

The manufacturer of a car battery specifies that it has a 30 ampere-hour capacity.
The term '30 ampere-hour' denotes the battery's

- **A** electrical energy content
- **B** available charge
- **C** electrical capacitance
- **D** life time in normal usage
- **E** electrical power

(OLE)

Solution 4.2

30 A h means the battery will provide 30 A for 1 h, or 10 A for 3 h or 2 A for 15 h, etc. In each case, the total charge that flows is $(1 \text{ A} \equiv 1 \text{ C s}^{-1})$

$$Q = (30 \text{ A})(3600 \text{ s}) = 108\,000 \text{ C}$$

Answer **B**

If the e.m.f. of the battery was known, e.g. 2.0 V, then the total energy content can be calculated. It is $(1 \text{ V} \equiv 1 \text{ J C}^{-1})$ $E = (2.0 \text{ V})(108\,000 \text{ C}) = 216\,000 \text{ J}$

Example 4.3

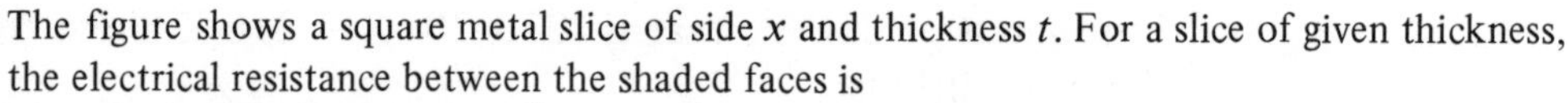

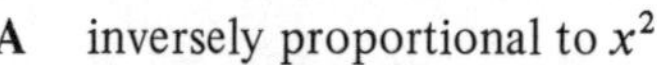

The figure shows a square metal slice of side x and thickness t. For a slice of given thickness, the electrical resistance between the shaded faces is

A inversely proportional to x^2
B inversely proportional to x
C independent of x
D proportional to x
E proportional to x^2 (NISEC)

Solution 4.3

Let the resistivity of the metal be ρ. The resistance of the slice between the shaded faces is

$$R = \rho \frac{x}{xt} = \frac{\rho}{t} \qquad \text{which is independent of } x$$

Answer **C**

An odd answer one wouldn't guess! Must it be a square sheet? Try a rectangular one x by $3x$.

Example 4.4

(a) Mains leads for electric fires consist of several strands of thin wire rather than a single thick conductor. Explain this design for the leads. [2]

(b) A cable for such a lead consists of 15 copper strands each of diameter 0.20 mm. Calculate the resistance of 6.0 m of this cable given that the resistivity of copper is 1.7×10^{-8} Ω m. If the cable had been made of a single copper wire, what would be its diameter in order that this cable had the same resistance per metre as the stranded cable? [8]

Solution 4.4

(a) The leads need to be flexible; thin wires bend with little strain and hence without plastic deformation.

A full explanation would be long and worth many more than the 2 marks which appear to be on offer here.

(b) For a single strand

$$R = \rho \frac{l}{A} = \frac{(1.7 \times 10^{-8}\ \Omega\ \text{m})\ (6.0\ \text{m})}{\pi\ (1.0 \times 10^{-4}\ \text{m})^2} = 3.25\ \Omega$$

$A = \pi r^2$, beware diameter.

With 15 strands in parallel, cable resistance $= \dfrac{3.25\ \Omega}{15} = 0.22\ \Omega$

A thick copper wire of radius r and length 6.0 m will have the same R, and hence the same resistance per metre, when

$$0.22\ \Omega = \frac{(1.7 \times 10^{-8}\ \Omega\ \text{m})\ (6.0\ \text{m})}{\pi r^2}$$

$$\Rightarrow \quad r = 3.84 \times 10^{-4}\ \text{m}$$

i.e. when its diameter is 0.77 mm.

Example 4.5

The graph shows I–V characteristics for an ohmic resistor R and a germanium diode D. (The reverse characteristic of the diode is not shown.)

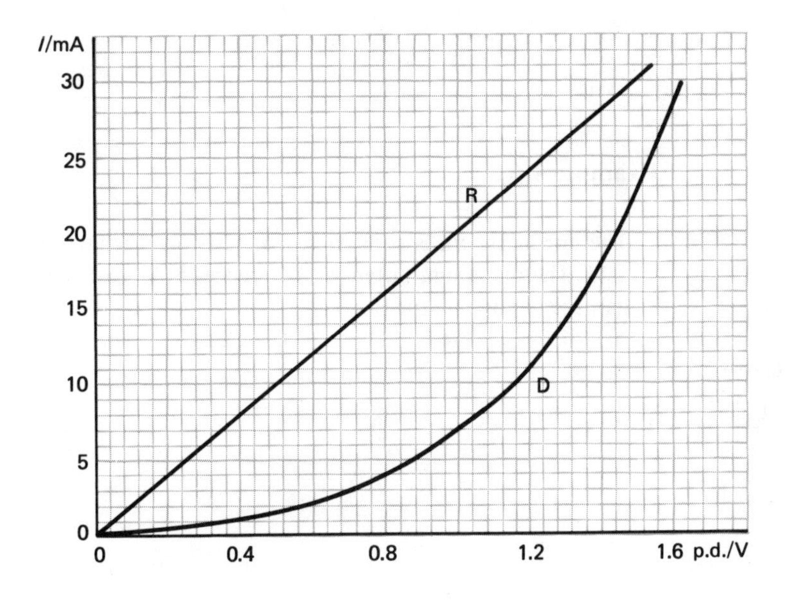

(a) Describe how the resistance of (i) the resistor and (ii) the diode vary as the p.d. across them increases.
(b) What is the current in a cell of e.m.f. 1.5 V when R and D are connected to it in parallel? (Assume the cell has zero internal resistance.)
(c) What is the p.d. needed to send a current of (i) 10 mA and (ii) 20 mA through R and D when they are connected in series?
(d) R and D are connected in series to a cell of zero resistance and of e.m.f. 2.0 V. Estimate the current in the circuit. [12]

Solution 4.5

(a) (i) R has a constant resistance of 50 Ω.

1.0 V, 20 mA ⇒ R = 1.0 V/0.02 A.

(ii) D has a resistance that varies. It gets smaller as the p.d. across it rises.

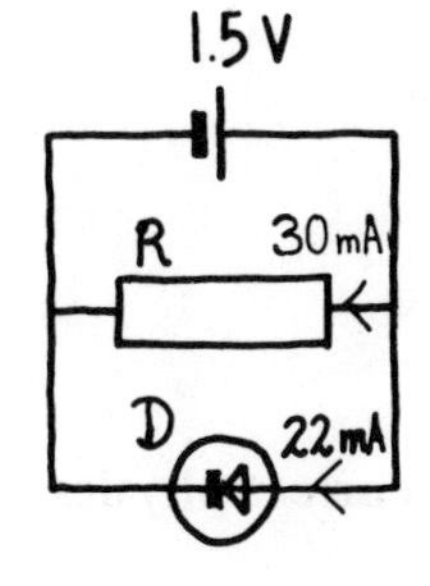

For example, at 5 mA, R_D = 176 Ω
at 10 mA, R_D = 115 Ω
at 20 mA, R_D = 72 Ω.

(b) Current from the cell = 30 mA + 22 mA
= 52 mA

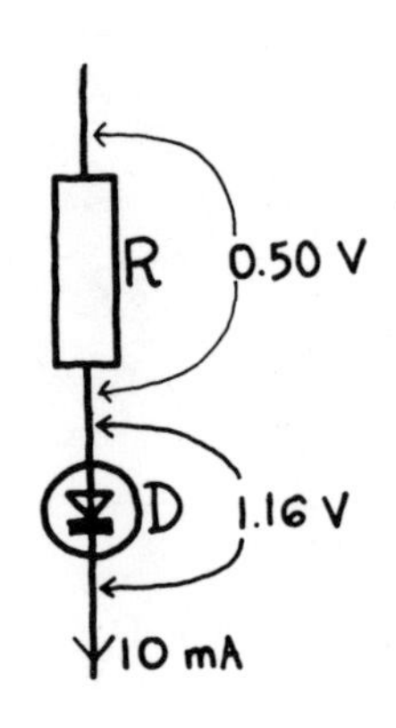

(c) (i) Required p.d. = 0.50 V + 1.16 V = 1.66 V
(ii) Similarly at 20 mA
required p.d. = 1.00 V + 1.45 V = 2.45 V

The readings of current and p.d. are taken straight from the graph. Calculations are not necessary for R and aren't possible for D.

(d) From (c) the current lies between 10 mA and 20 mA. The current for which $V_R + V_D$ is 2.0 V is approximately 14 mA.

This is difficult as you can't calculate how the p.d. splits between R and D.

Example 4.6

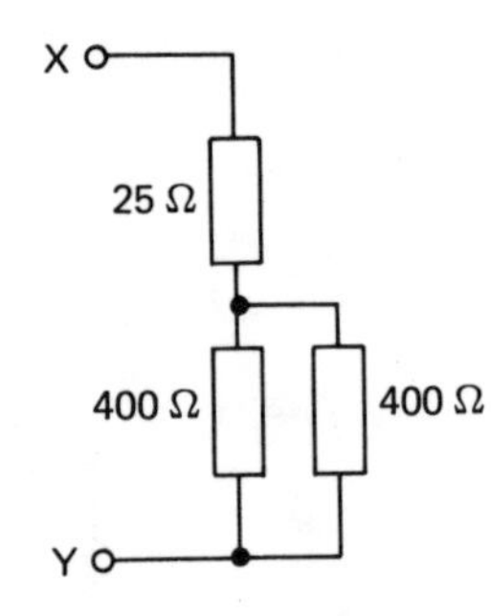

The circuit shown is constructed of resistors, each of which has a maximum power rating of 1 W. The maximum potential difference, in V, that can safely be applied across XY without damage to any resistor is

A 11.3 **B** 21.3 **C** 22.5 **D** 42.5 **E** 45.0 (NISEC)

Solution 4.6

The two 400 Ω resistors add to be 200 Ω. Thus a p.d. V across XY will split in the ratio 25:200 giving a p.d. of

$$\frac{25\ \text{V}}{25 + 200} = \frac{V}{9}$$

across the 25 Ω resistor.

The power condition for the 25 Ω resistor is

$$P_{max} = 1\ \text{W} > \left(\frac{V}{9}\right)^2 \div 25\ \Omega$$

$$V^2 < (81\ \text{W})\,(25\ \Omega) \quad \Rightarrow \quad V < 45\ \text{V}$$

The answer is E if $8V/9 = 40$ V does not result in more than 1 W in a 400 Ω resistor. But $P = (40\ \text{V})^2/400\ \Omega = 4$ W, so E is wrong.

The power condition for the 400 Ω resistors is

$$P_{max} = 1\ \text{W} > \left(\frac{8V}{9}\right)^2 \div 400\ \Omega$$

$$V^2 < \left(\frac{81}{64}\ \text{W}\right)(400\ \Omega) \quad \Rightarrow \quad V < 22.5\ \text{V}$$

Answer **C**

Note how the wrong responses, E in this case, can be calculated by correct physics.

Example 4.7

What must be the values of R_1 and R_2 for the lamps A and B to operate at their stated ratings? The 6 V battery can be assumed to have zero internal resistance. [7]

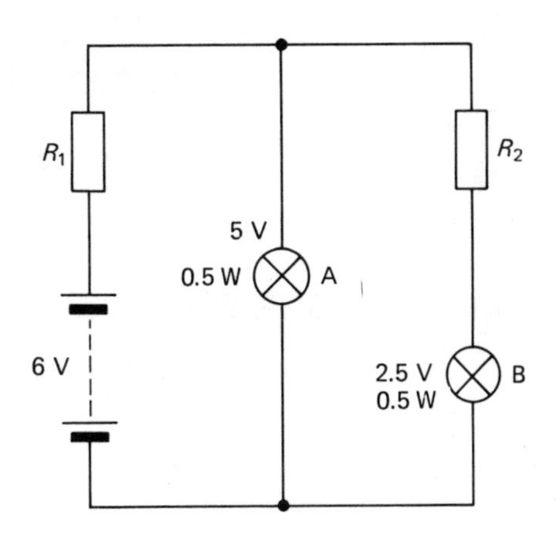

Solution 4.7

Current in A ($P = IV$) = 0.1 A
Current in B = 0.2 A
Kirchhoff's first law, current in R_1 = 0.3 A
From Kirchhoff's second law

p.d. across R_2 = 5 V – 2.5 V = 2.5 V
p.d. across R_1 = 6 V – 5 V = 1 V

$$\therefore R_1 = \frac{1\ \text{V}}{0.3\ \text{A}} = 3.3\ \Omega$$

$$R_2 = \frac{2.5\ \text{V}}{0.2\ \text{A}} = 12.5\ \Omega$$

Example 4.8

The diagram shows a potential divider P of resistance 10 000 Ω connected to a 10.0 V supply of negligible internal resistance. One terminal of the voltmeter of resistance 2000 Ω is connected to the mid point of the divider. What will the voltmeter reading be? Explain why the value of the voltage would be different if it were measured using a potentiometer rather than a voltmeter. [7]

(AEB 1983)

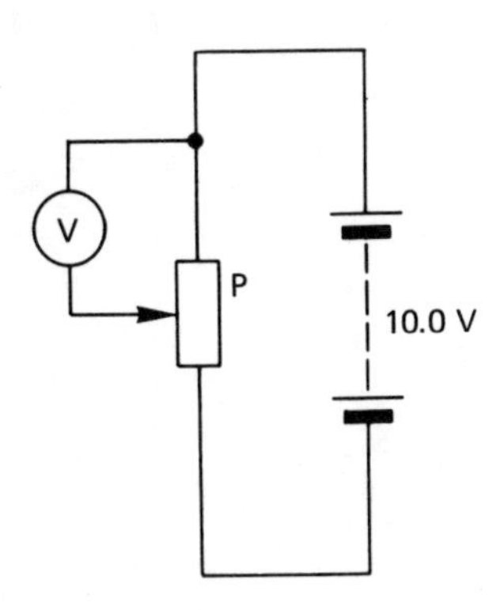

Solution 4.8

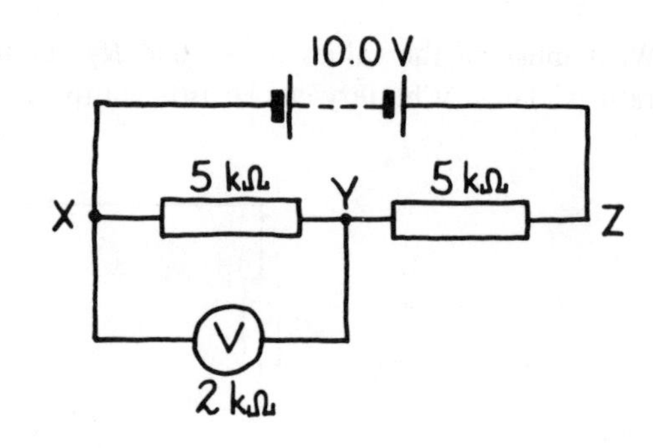

Net resistance R of 5000 Ω and 2000 Ω connected in parallel is given by

$$\frac{1}{R} = \frac{1}{5000\ \Omega} + \frac{1}{2000\ \Omega}$$

$$\Rightarrow \quad R = 1430\ \Omega$$

So the total resistance of the circuit = 5000 Ω + 1430 Ω = 6430 Ω

$$\text{Current in battery} = \frac{10.0\ \text{V}}{6430\ \Omega} = 0.0156\ \text{A}$$

p.d. between Y and Z = (0.0156 A) (5000 Ω) = 7.8 V
p.d. between X and Y = 10.0 V − 7.8 V = 2.2 V
This will be the reading of the voltmeter.

No current is drawn by a potentiometer at the moment the readings are taken. As it does not affect the circuit it can be thought of as having infinite resistance. The voltage would now be 5.0 V.

Example 4.9

A newly purchased dry cell, labelled $1\frac{1}{2}$ V, is connected in series with a switch and a fixed resistor of resistance 1.00 Ω. A very high resistance voltmeter is connected across the resistor.

The switch is closed at time $t = 0$ and the potential difference, V, across the resistor is measured every 20 s, including a period after the switch is opened at time $t = 100$ s.

(a) What is the e.m.f. of the cell at $t = 0$? Calculate the internal resistance, r, of the cell at this instant, i.e. at $t = 0$. [4]
(b) Calculate the internal resistance, r', of the cell at $t = 100$ s. [3]
(c) Draw a circuit diagram showing how a slide-wire potentiometer can be used to measure the potential difference across the terminals of a dry cell so as to produce the graph in (a). Explain how the experiment would have been performed. [7]

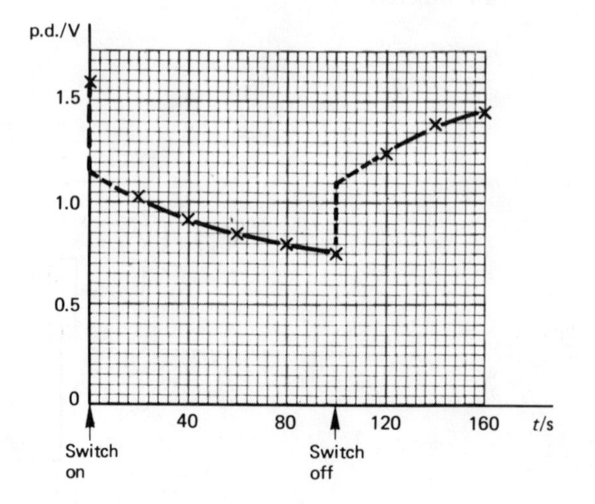

Solution 4.9

(a) At $t = 0$, the e.m.f. is 1.60 V.
At the moment the switch is closed, the current in the 1.00 Ω resistor is

$$I = V/R = 1.15\ \text{V}/1.00\ \Omega = 1.15\ \text{A}$$

so, using $E = I(R + r)$

$$1.60\ \text{V} = (1.15\ \text{A})\,(1.00\ \Omega + r) \Rightarrow r = 0.39\ \Omega$$

(b) At $t = 100$ s, the current in the 1.00 Ω resistor is 0.75 V/1.00 Ω = 0.75 A and, as the e.m.f. is 1.10 V at the moment the switch is opened then

$$1.10\ \text{V} = (0.75\ \text{A})\,(1.00\ \Omega + r') \Rightarrow r' = 0.47\ \Omega$$

The e.m.f. of the cell varies during the experiment owing to polarisation - chemical effects.

(c)

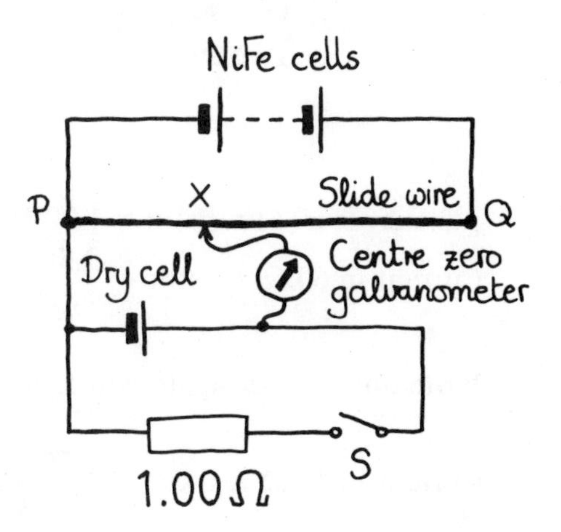

To calibrate the potentiometer a known p.d. must be connected across PX. To do this open the switch S and replace the dry cell with a standard cell of known e.m.f., E_s, and find a balance point PX = l_s. With the dry cell in place, close S and take a series of readings of the time, t, and of the balance lengths, l, every 20 s, opening S after 100 s. For each balance reading $V = (l/l_s)E_s$.

Example 4.10

The 4.00 V cell in the circuits shown below has zero internal resistance.

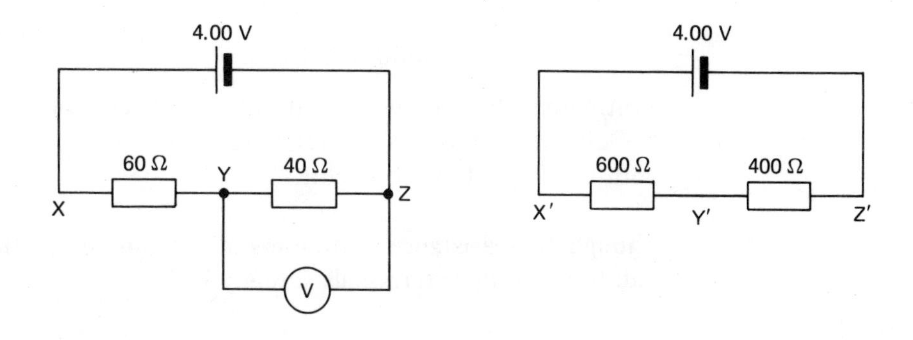

An accurately calibrated voltmeter connected across YZ records 1.50 V. Calculate
(a) the resistance of the voltmeter,
(b) the voltmeter reading when it is connected across Y′Z′.
What do your results suggest concerning the use of voltmeters? [5]
(L)

Solution 4.10

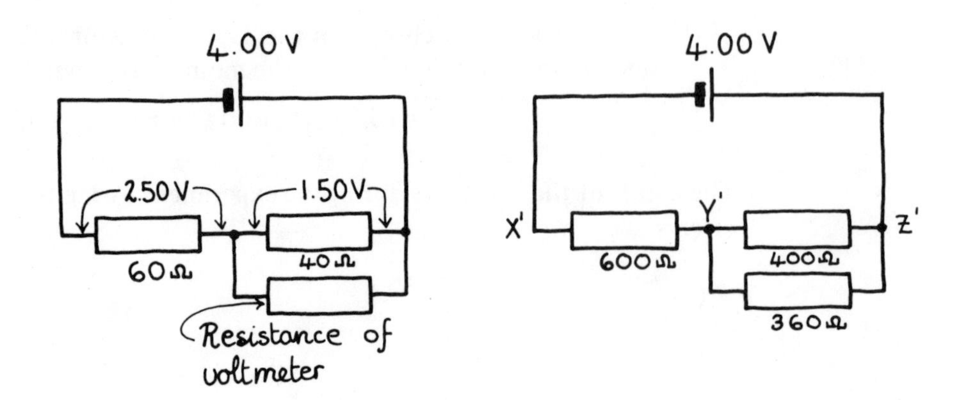

(a) p.d. across 60 Ω resistor = 4.00 V – 1.50 V = 2.50 V

Current in 60 Ω resistor = $\dfrac{2.50\ \text{V}}{60\ \Omega}$ = 0.0417 A

Write down extra significant figures if a subtraction is imminent.

Current in 40 Ω resistor = $\dfrac{1.50\ \text{V}}{40\ \Omega}$ = 0.0375 A

Current in voltmeter = 0.0417 A – 0.0375 A = 0.0042 A

An ideal voltmeter would draw no current.

Resistance of voltmeter = $\dfrac{1.50\ \text{V}}{0.0042\ \text{A}}$ = 360 Ω

If time is available it is worth checking backwards through the calculation.

(b) If the resistance between Y′ and Z′ is R then

$$\frac{1}{R} = \frac{1}{400\ \Omega} + \frac{1}{360\ \Omega} \qquad \text{so} \qquad R = 189\ \Omega$$

Let p.d. between X′ and Y′ with voltmeter in place be V,

$$V = \left(\frac{189\ \Omega}{600\ \Omega + 189\ \Omega}\right)(4.00\ \text{V}) = 0.96\ \text{V}$$

Voltmeters should have an internal resistance large compared with that in the circuit so that the current drawn by the voltmeter is small compared with the current in the circuit.

Though low-resistance voltmeters affect the circuit they do read what the p.d. between their terminals actually is.

Example 4.11

A high voltage power supply is known to have an e.m.f. of 2000 V. But when a voltmeter of resistance 10 kΩ is connected to the output terminals of the supply, a reading of only 2 V is obtained.

(a) Explain this observation.

(b) Find

(i) the current in the meter;

(ii) the internal resistance of the power supply. [8]

Solution 4.11

(a) As the p.d.'s round the circuit must add up to 2000 V, the power supply must have a high internal resistance across which the 1998 V is dropped.

High-voltage power supplies are not generally designed to supply anything but tiny currents.

(b) (i) Voltmeters read the p.d. between their terminals so the p.d. across the power supply really is 2 V.

$$\text{Current in voltmeter} = \frac{2\text{ V}}{10\,000\ \Omega} = 0.0002\text{ A}$$

(ii) Internal resistance of the power supply $= \dfrac{1998\text{ V}}{0.0002\text{ A}}$

$$= 1 \times 10^7\ \Omega$$

Example 4.12

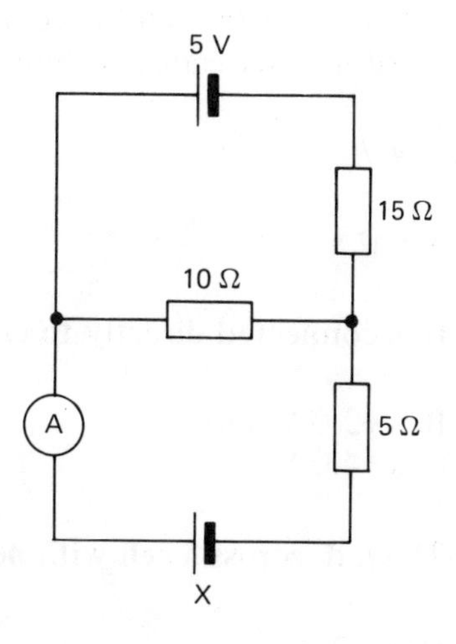

When the circuit shown is set up, it is found that the current in the ammeter is zero. Which of the following deductions can be made?

1 The e.m.f. of cell X is 2 V.

2 The potential difference across the 5 Ω resistor is 3 V.

3 The total power dissipated in the circuit is 1 W. (NISEC)

Solution 4.12

With no current in A the upper part of the circuit is a simple circuit, the 5 V splits to produce a p.d. of 3 V across the 15 Ω and 2 V across the 10 Ω resistor.
The e.m.f. of cell X = p.d. across the 10 Ω = 2 V
(1 is correct)

There is no current in the 5 Ω resistor so there can be no p.d. across it.
(2 is incorrect)

The power dissipated is the result of 5 V across a resistance of 25 Ω, i.e.

$P = V^2/R = (5\ \text{V})^2/(25\ \Omega) = 1\ \text{W}$

(3 is correct)
Answer 1 ✓ 2 ✗ 3 ✓

Example 4.13

Each of the cells in the circuit below has an e.m.f. of 2.0 V and negligible internal resistance. The resistors are both 1.0 kΩ. The earth connection is at 0 V.

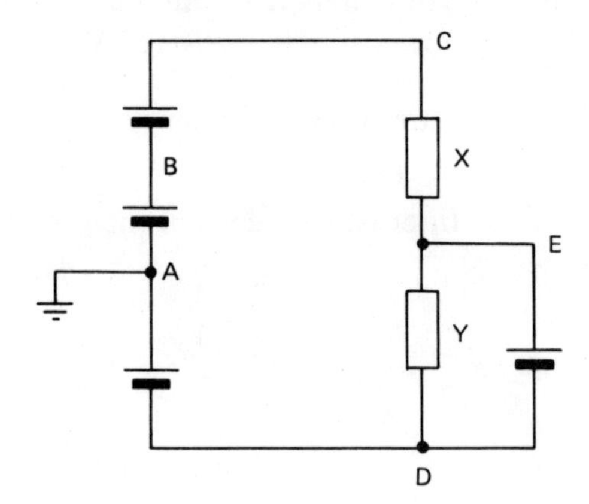

(a) Write down the potentials of the points A to E. [3]
(b) Calculate the current in resistor X. [2]

Solution 4.13

(a) A is 0 V.

It is connected directly to earth.

B is +2.0 V and
C is +4.0 V

The p.d. across a cell with negligible internal resistance is its e.m.f.

D is –2.0 V

Because of the cell its potential is 2.0 V below earth.

E is 0 V

The resistor makes no difference. Ohm's law has not been used so far.

(b) p.d. across X = 4.0 V − 0 V = 4.0 V

Therefore current in X = $\dfrac{4.0\ \text{V}}{1000\ \Omega} = 4 \times 10^{-3}$ A

Example 4.14

To investigate the characteristics of a power supply unit a girl uses an ammeter and a resistance box in the following circuit.

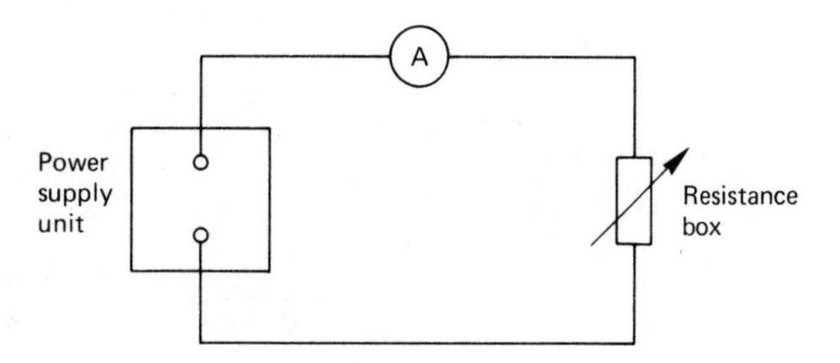

(a) For this circuit derive the expression

$$R = \frac{E}{I} - r$$

where E is the e.m.f. of the power supply unit, r is its internal resistance, R is the total external resistance and I is the current in the circuit.

She collects values of I and R and draws the following graph.

(b) Explain why she chooses to draw a graph of R against $1/I$.

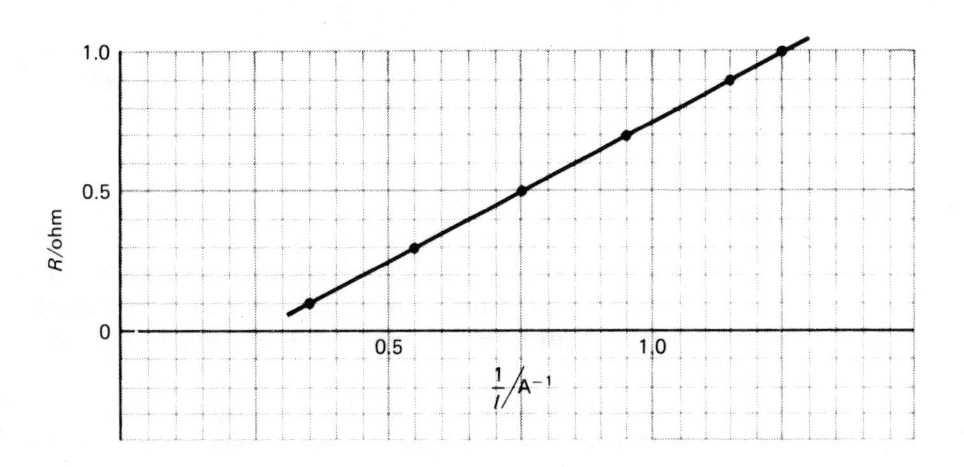

(c) Extend the graph and deduce from it:
   (A) the maximum current if the power supply were short circuited;
   (B) the internal resistance of the power supply unit;
   (C) the e.m.f. of the power supply unit.

(d) Under what conditions will this power supply unit transfer maximum power to an external circuit?

(e) Calculate the maximum power it can transfer.

(f) State another method by which she could check the value of the e.m.f. of the power supply unit. [10]

(SEB)

Solution 4.14

(a) $E = I(R + r)$ and therefore $R + r = E/I$
So $R = E/I - r$ which is what had to be derived.

(b) The general formula for a straight line graph is:

$y = mx + c$, m and c are constants

If x is $1/I$, m is E and c is $-r$ then y is R and it will be straight line.

(c) (A) Short circuit means that the external resistance R is zero and so the x intercept gives that $1/I = 0.25\ \text{A}^{-1}$

$$\text{Maximum current} = \frac{1}{0.25\ \text{A}^{-1}} = 4\ \text{A}$$

(B) When $1/I = 0$, $R = -r$. The y intercept is -0.24, so $r = 0.24\ \Omega$.

(C) The gradient of the graph is E so:

$$E = \frac{y_2 - y_1}{x_2 - x_1} = \frac{0.5\ \Omega - 0\ \Omega}{0.70\ \text{A}^{-1} - 0.25\ \text{A}^{-1}} = \frac{0.5\ \Omega}{0.5\ \text{A}^{-1}} = 1\ \text{V}$$

(d) Maximum power transfer to the external circuit occurs when the external resistance is the same as the internal resistance. So $R = 0.24\ \Omega$ for maximum power transfer.

(e) Power in a resistor $= I^2R$
From the graph $1/I = 0.5\ \text{A}^{-1}$ when $R = 0.24\ \Omega$, so $I = 2.0\ \text{A}$
Power $= (2.0\ \text{A})^2\ (0.24\ \Omega) = 0.96\ \text{W}$

(f) The p.d. across the terminals is the e.m.f. when it is measured with a high-resistance voltmeter. That is one that draws very little current.

Example 4.15

(a) Explain what is meant by electrical resistance. [2]

(b) The temperature coefficient of resistance of tungsten is $4.8 \times 10^{-3}\ \text{K}^{-1}$. Calculate the resistance at room temperature, 27°C, of a tungsten filament lamp rated 100 W 240 V if it has a working temperature of 2600 K. [6]

Solution 4.15

(a) If, when a p.d. is applied across an electrical conductor, a current I flows through it, the conductor is said to have a resistance calculated as V/I; it is measured in ohms. For metal conductors it describes how difficult it is for the conduction electrons to move through the lattice ions.

This is little more than the defining equation $R = V/I$, but something extra is required by the phrase 'what is meant by'. See also Example 5.5.

(b) $R = R_0\ (1 + \alpha\Delta\theta)$ where $\alpha = 4.8 \times 10^{-3}\ \text{K}^{-1}$
At 2600 K, $P = V^2/R$

$$\therefore \quad R = \frac{V^2}{P} = \frac{(240\ \text{V})^2}{100\ \text{W}} = 576\ \Omega$$

So at 27°C, which is 300 K, the resistance R_0 is given by

$$576\ \Omega = R_0\,[1 + (4.8 \times 10^{-3}\ \text{K}^{-1})\,(2300\ \text{K})]$$

$$= R_0\,[1 + 11]$$

$$\text{i.e.} \quad R_0 = 576\ \Omega/12 = 48\ \Omega$$

Example 4.16

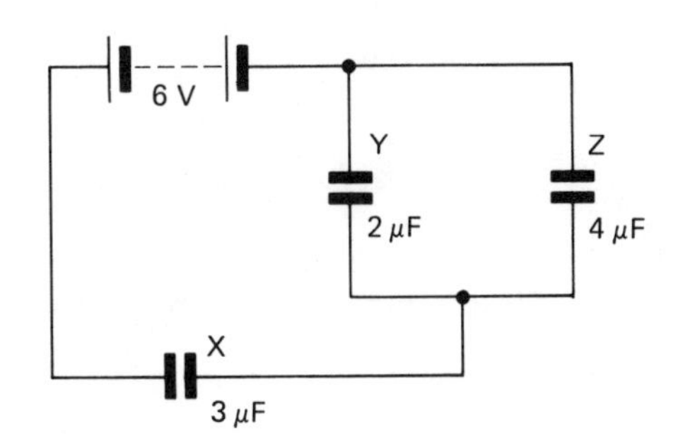

Examine the circuit and calculate
(a) the potential difference across capacitor X,
(b) the charge on the plates of capacitor Y,
(c) the energy associated with the charge stored in capacitor Z. [6]

Solution 4.16

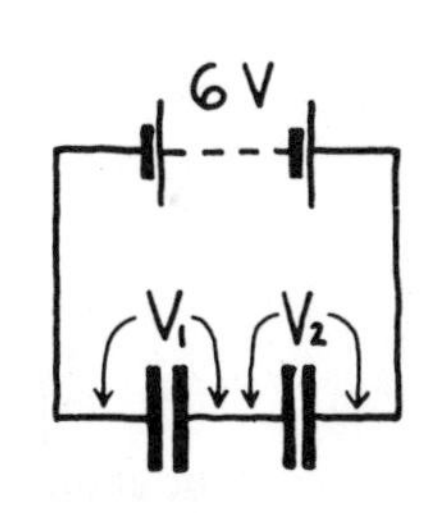

(a) The sum of the charges stored on Y and Z must be the same as the charge Q stored on X, because the charge on the right plate of X can only come from the bottom plates of Y and Z.
The combined capacitance of Y and Z = 2 μF + 4 μF = 6 μF

$Q = CV = (3\ \mu F)V_1 = (6\ \mu F)V_2$ i.e. $V_1 = 2V_2$

Also $V_1 + V_2 = 6$ V so that clearly $V_1 = 4$ V and $V_2 = 2$ V
The p.d. across X is therefore 4 V.
(b) Charge on Y is $CV = (2 \times 10^{-6}\ F)\ (2\ V) = 4 \times 10^{-6}$ C.
(c) The energy stored on Z is $\frac{1}{2}CV^2 = (0.5)\ (4 \times 10^{-6}\ F)\ (2\ V)^2$
$= 8 \times 10^{-6}$ J

It is important in questions like this to stick to the first principles of charge and potential, and not to be tempted into using results that may apply only to resistors.

Example 4.17

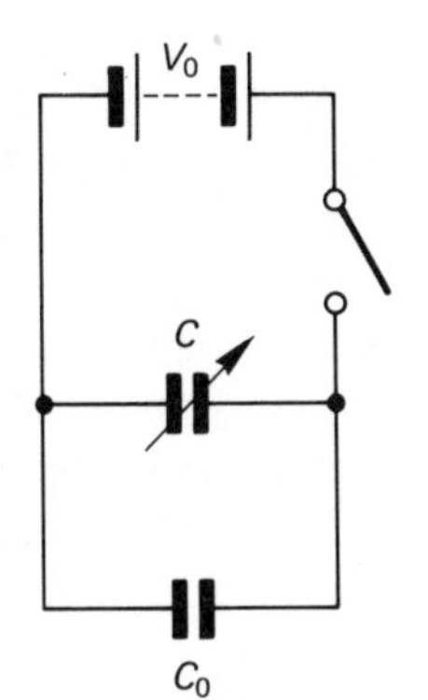

A fixed capacitor of capacitance C_0, and a variable capacitor of capacitance C are connected as shown to a battery of e.m.f. V_0.
(a) Initially the switch is closed and C is adjusted so that $C = C_0$. Write down in terms of C_0 and V_0 the total energy stored in the two capacitors.
(b) The switch is then opened and C is adjusted so that $C = 3C_0$.
(i) What is the resulting potential difference across the capacitors?
(ii) Calculate the total energy now stored in the two capacitors.
(iii) Explain the difference in the stored energies. [7]

Solution 4.17

(a) Each capacitor stores energy $= \frac{1}{2}C_0 V_0^2$
i.e. total energy stored $= C_0 V_0^2$.
(b) (i) Before C is altered each capacitor has charges $\pm Q_0 = C_0 V_0$ on its plates. With the switch open this charge cannot escape so that the

total charge on the two capacitors remains $2Q_0 = 2C_0 V_0$.

The new equivalent capacitance, of C_0 in parallel with $3C_0$, is $4C_0$.

$$\therefore \text{ New p.d.} \qquad V = \frac{2C_0 V_0}{4C_0} = \frac{V_0}{2}$$

(ii) The p.d. across each capacitor is now $\dfrac{V_0}{2}$

$$\text{New total energy stored } = \tfrac{1}{2}C_0 \left(\frac{V_0}{2}\right)^2 + \tfrac{1}{2} \times 3C_0 \left(\frac{V_0}{2}\right)^2$$

$$= \tfrac{1}{2}C_0 V_0^2$$

(iii) The energy lost has become internal energy in the wires connecting the two capacitors as charge flows through them during the increase in capacitance of C.

As well as the loss of stored energy there is work done as C increases, e.g. by letting the plates come closer together. This energy is also converted to internal energy in the wires.

Example 4.18

(a) Define the term capacitance as applied to a capacitor. [2]

(b) Describe an experiment to measure the capacitance of a capacitor known to be about 10 μF. [8]

(c) A capacitor forming part of a computer memory-store consists of a pair of rectangular conducting strips each measuring 30×10^{-6} m by 40×10^{-6} m separated by an oxide film of thickness 1.4×10^{-7} m. The resistivity of the oxide is 1.5×10^{19} Ω m and its relative permittivity is 4.0. Initially one strip is charged to a potential of +5.0 V, the other being permanently at zero potential. The device is found to discharge very slowly through the leakage resistance of the oxide film; other leakage paths are negligible.

[Take the permittivity of vacuum, ϵ_0, to be 8.8×10^{-12} F m^{-1}, and the electronic charge, e, to be -1.6×10^{-19} C.]

(i) Calculate the capacitance of this device. [4]

(ii) Calculate the resistance of the leakage path. [4]

(iii) Hence, calculate the time elapsing before the potential of the charged strip falls to +3.0 V, assuming that the mean rate of loss of charge is equal to that corresponding to a mean potential of +4.0 V. [6]

(iv) What is the initial rate of leakage through the device in terms of the rate of flow of electrons? [4]

(v) How would the initial rate of leakage compare with that in another device identical in all respects except that the oxide film was twice as thick? [2]

(OLE)

Solution 4.18

(a) A capacitor is two conductors separated by an insulator. The conductors carry equal, but opposite, charges $+Q$ and $-Q$. Q is proportional to V, the potential difference between the conductors and the capacitance, C, is defined from the equation: $Q = CV$.

(b) A reed switch technique.

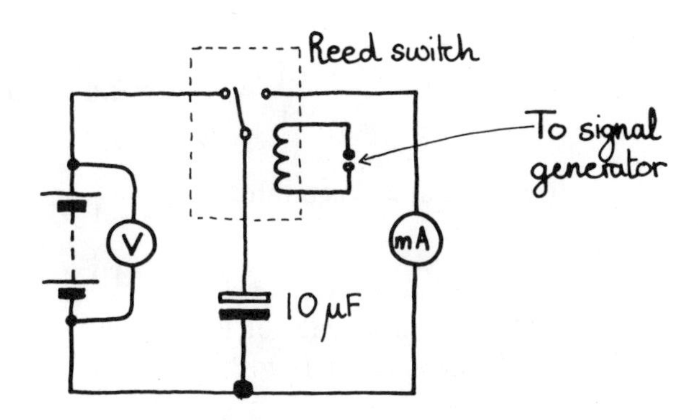

A current from the signal generator of frequency about 200 Hz operates the switch and causes the capacitor to be repeatedly charged and then discharged through the milliammeter.

The capacitance is calculated from the average discharge current, I, the frequency of the reed switch, f, and the p.d. across the capacitor when it is fully charged.

Charge through meter each second = $fQ = fCV$, which is I.

Therefore $\quad C = \dfrac{I}{fV}$

For an accurate result one would plot a graph of current against frequency. The gradient would be CV.

It is important to be sure that the capacitor is fully discharged each time. A levelling off in the graph would show that the RC time for the discharge was not short compared with the period of the charge–discharge cycle. The resistance of the milliammeter might be 50 Ω.

$RC = (50\ \Omega)\ (10^{-5}\ \text{F}) = 5 \times 10^{-4}\ \text{s}$

This is well below the time of about 5×10^{-3} s for the cycle at a frequency of 200 Hz.

If V is about 10 V then

$I = (10^{-5}\ \text{F})\ (200\ \text{Hz})\ (10\ \text{V})$
$= 2 \times 10^{-2}\ \text{A} = 20\ \text{mA}$

It is clear from this that a milliammeter is required.

(c) (i) For a parallel-plate capacitor

$$C = \frac{\epsilon_r \epsilon_o A}{d}$$

$$C = \frac{(4.0)\ (8.8 \times 10^{-12}\ \text{F m}^{-1})\ (30 \times 10^{-6}\ \text{m})\ (40 \times 10^{-6}\ \text{m})}{(1.4 \times 10^{-7}\ \text{m})}$$

$$= 3.0 \times 10^{-13}\ \text{F}$$

(ii) $R = \dfrac{\rho l}{A} = \dfrac{(1.5 \times 10^{19}\ \Omega\ \text{m})\ (1.4 \times 10^{-7})}{(30 \times 10^{-6}\ \text{m})\ (40 \times 10^{-6}\ \text{m})} = 1.8 \times 10^{21}\ \Omega$

Note that the current path is across the capacitor.

(iii) Mean leakage current = $\dfrac{4.0\ \text{V}}{1.8 \times 10^{21}\ \Omega} = 2.3 \times 10^{-21}\ \text{A}$

Charge lost as p.d. falls from 5.0 V to 3.0 V is ΔQ, where

$\Delta Q = C\Delta V = (3.0 \times 10^{-13}\ \text{F})\ (2.0\ \text{V}) = 6.0 \times 10^{-13}\ \text{C}$
Since $Q = It$,

$$t = \frac{Q}{I} = \frac{6.0 \times 10^{-13}\ \text{C}}{2.2 \times 10^{-21}\ \text{A}} = 2.6 \times 10^{8}\ \text{s}$$

(iv) Initial leakage rate $= \dfrac{5.0\ \text{V}}{(1.8 \times 10^{21}\ \Omega)\ (1.6 \times 10^{-19}\ \text{C/el})}$

$= 1.7 \times 10^{-2}$ electrons per second

(v) The initial rate of leakage does not depend on the value of C, so it is halved because with the thicker oxide layer R is doubled.

Example 4.19

A student designs the following circuit in an attempt to measure the speed of a rifle bullet. A and B are two pieces of foil which the bullet breaks before embedding itself in a wooden block (not shown) placed behind B. The R-C part of the circuit is used as a clock.

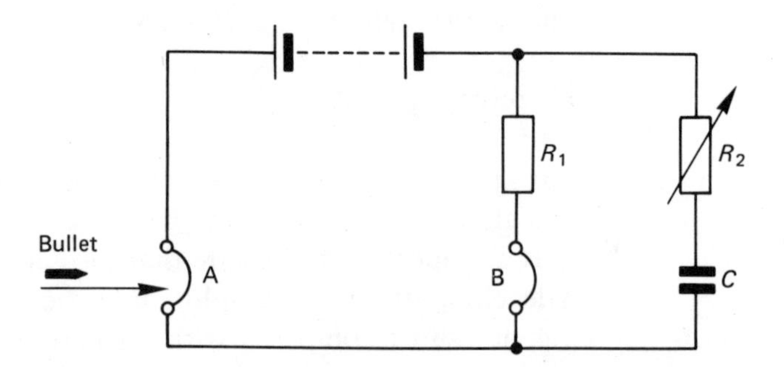

(a) Explain what measurements you would take in order to calculate the speed of the bullet. [4]
(b) For a bullet from a high-velocity rifle which is expected to have a speed of about 200 m s^{-1}, suggest a suitable value for R_2 if C is a 0.47 μF capacitor and R_1 is a fixed resistor of 2.0 kΩ. Take the distance AB to be 0.4 m. [4]
(c) Suggest how the experiment might be developed to provide a more reliable result for the bullet's speed. [2]

Solution 4.19

(a) The speed of the bullet v = AB/t, where t is the time between the bullet breaking the foil at A and then at B. AB can be measured with a ruler; t can be found by measuring the p.d. across the capacitor C (i) before A is broken, and (ii) after B is broken. This must be done with a very high resistance voltmeter, e.g. a digital voltmeter.

During the flight of the bullet from A to B the capacitor discharges exponentially through R_1 and R_2, the discharge starts at the moment A is broken. You are not asked to explain this but you must understand it in order to answer the question.
If the readings are (i) V_0 and (ii) V then

$$V = V_0\ \text{e}^{-t/RC}$$

$$\frac{t}{RC} = \ln\left(\frac{V_0}{V}\right)$$

so that $t = RC \ln (V_0/V)$

(b) The bullet takes $t = 0.4\ \text{m}/200\ \text{m s}^{-1} = 0.002$ s to cross from A to B, so the time constant for the discharge circuit, the time for the capacitor to discharge to about $\frac{1}{3}$ of its initial p.d., should be arranged to be about this time.

If $(R_1 + R_2)C = 0.002$ s

then $$R_2 + 2000\ \Omega = \frac{0.002\ \text{s}}{0.47 \times 10^{-6}\ \text{F}} = 4300\ \Omega$$

$$\Rightarrow R_2 = 2300\ \Omega$$

(c) The distance AB could be varied and, using appropriate values for R_2 each time, a graph of the distance AB against t plotted. The slope of the graph represents the bullet's speed.

Designing an experiment which is to yield reliable results usually involves altering the variables so as to average out uncertainties in the measured values.

Example 4.20

When an insulated metal container is being filled with a liquid which has become charged by flowing along the input pipe, the container may be raised to a high potential. The magnitude of this potential depends on the rate at which charge is received from the liquid and the rate at which charge leaks to earth.

The charges may be considered to move in a circuit equivalent to that shown.

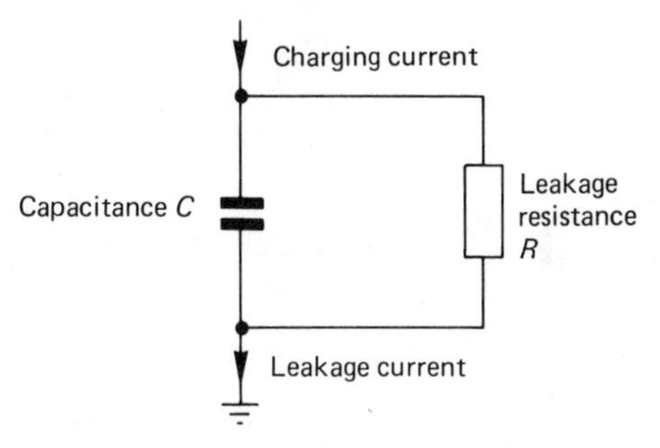

The liquid carries a charge of 0.2 microcoulombs per cubic metre and fills the container at a rate of 0.05 cubic metres per second. The capacitance of the container may be considered to be 60 picofarads and the leakage resistance between the container and earth to be 5×10^{12} ohms.

(a) Calculate the magnitude of the charging current.
(b) What is the maximum potential to which the container could be raised?
(c) What is the maximum electrical energy stored in the container?
(d) Why is this situation potentially dangerous? Suggest two ways of minimising this hazard.

[8]
(SEB)

Solution 4.20

(a) Charge flow is $0.2 \times 10^{-6}\ \text{C m}^{-3}$ at a rate of $0.05\ \text{m}^3\ \text{s}^{-1}$,

i.e. charging current $= \left(0.2 \times 10^{-6}\ \frac{\text{C}}{\text{m}^3}\right)\left(0.05\ \frac{\text{m}^3}{\text{s}}\right)$

$$= 1.0 \times 10^{-8}\ \frac{\text{C}}{\text{s}}$$

(b) As the capacitor charges, the p.d. V across it and the leakage resistor increases to a stage when the leakage current through R is equal to the charging current, i.e. when

$$V = (1.0 \times 10^{-8}\ \text{A})\ (5 \times 10^{12}\ \Omega)$$
$$= 5 \times 10^{4}\ \text{V}$$

(c) The energy stored $= \frac{1}{2}QV = \frac{1}{2}CV^2$

$$\Rightarrow \quad E = \tfrac{1}{2}(60 \times 10^{-12}\ \text{F})\ (5 \times 10^{4}\ \text{V})^2$$
$$= 0.075\ \text{J}$$

The energy needed to produce this dangerous situation appears to be very small, 0.075 J. But think of the product IVt when, although V is high, t is very small, perhaps a few microseconds.

(d) This energy may be released in the form of a spark thus igniting an inflammable vapour, from, e.g., flowing petrol.

To reduce the hazard either connect the pipe to the container with a conductor, or earth the container.

Dry air breaks down and sparks occur when the electric field is about 3×10^{6} V m^{-1}, a gap of 16 mm or so for the p.d. calculated in (b).

Example 4.21

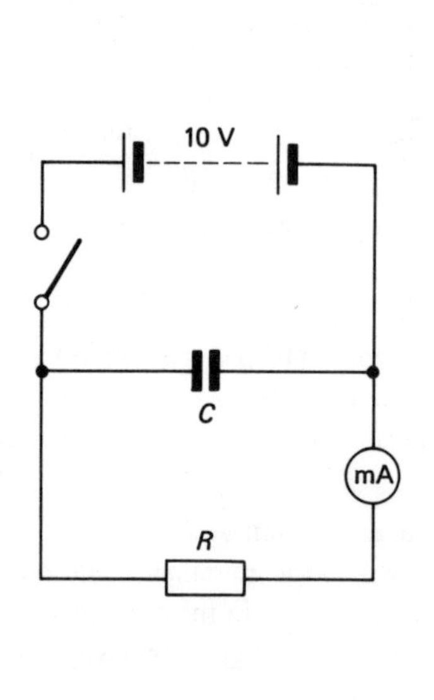

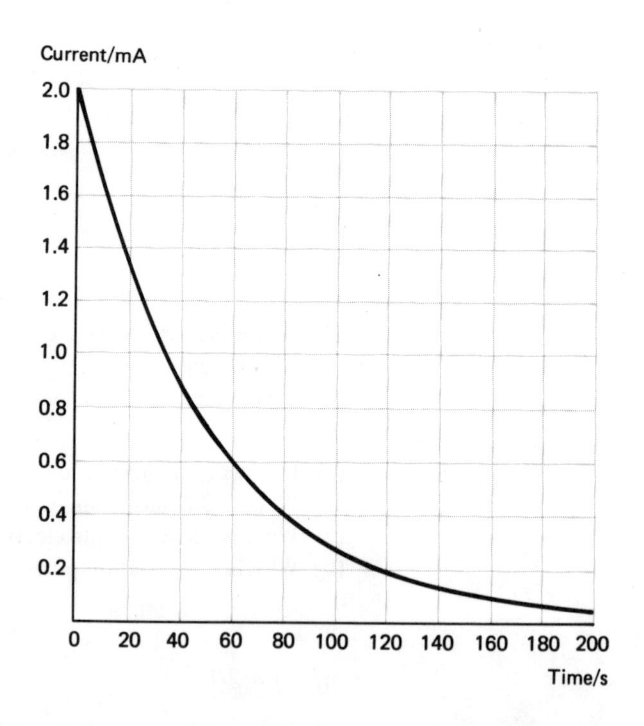

In an experiment to investigate the discharge of a capacitor through a resistor, the circuit shown in the diagram was set up. The battery had an e.m.f. of 10 V and negligible internal resistance. The switch was first closed and the capacitor allowed to charge fully. The switch was then opened (at time $t = 0$), and the graph shows how the milliammeter reading subsequently changed with time.

(a) Use your graph to estimate the initial charge on the capacitor. Explain how you arrived at your answer.

(b) Use your answer to (a) to estimate the capacitance of C.

(c) Calculate the resistance of R. [8]

(AEB 1984)

Solution 4.21

(a) The area under a current against time graph represents the charge that has flowed.

A square 0.2 mA by 20 s represents 4.0 mC.

There are about 25 squares of which half or more are under the graph, so an estimate of the charge would be:

$$\text{Charge} \approx 25 \times 4.0 \text{ mC} \approx 10^{-1} \text{ C}$$

You should only put one significant figure down in such an estimate.

(b) $\text{Capacitance} = \dfrac{\text{charge}}{\text{p.d.}} = \dfrac{10^{-1} \text{ C}}{10 \text{ V}} = 10^{-2} \text{ F}$

(c) The initial discharge current is 2 mA and the initial p.d. across the capacitor is 10 V. Since $R = V/I$

$$R = \frac{10 \text{ V}}{0.002 \text{ A}} = 5000 \text{ } \Omega$$

There is another way of getting (b) if you know (c). 1/e of the initial value of the current is 0.74 mA. From the graph you can find the time taken for the current to fall to this value. It is 50 s and this time is *RC*.

Therefore $C = \dfrac{50 \text{ s}}{5000 \text{ } \Omega} = 1.0 \times 10^{-2} \text{ F}$

The question has more information than is necessary so it is possible to check your calculation if you have time.

Example 4.22

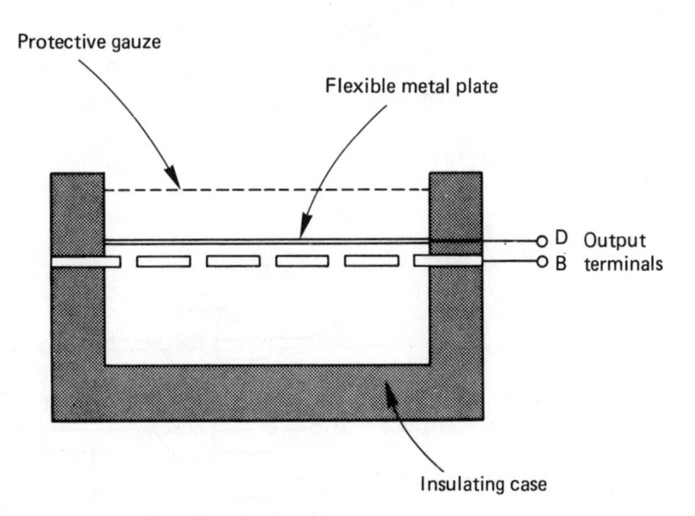

The diagram shows a section through a capacitor microphone. Its terminals B and D are connected in series with a resistor of 200 MΩ and a d.c. source of e.m.f. 100 V. The flexible plate D, which has an area 4 cm^2, oscillates when a sound wave strikes it.

(a) If the capacitance of the microphone is 20 pF calculate the separation of D and B.

(b) Calculate the time constant of the circuit and explain what will happen to (i) the capacitance of, and (ii) the charge on the microphone when a sound wave of high frequency, over 5000 Hz, strikes it.

(c) Hence explain why the p.d. across the 200 MΩ resistor oscillates at the same frequency as the sound wave.

(Take ϵ_0 to be 9×10^{-12} F m^{-1}.) [8]

Solution 4.22

(a) $C = \epsilon_0 \dfrac{A}{d} \quad \Rightarrow \quad d = \dfrac{\epsilon_0 A}{C}$

i.e. $d = \dfrac{(9 \times 10^{-12}\ \text{F m}^{-1})(4 \times 10^{-4}\ \text{m}^2)}{20 \times 10^{-12}\ \text{F}}$

$= 1.8 \times 10^{-4}$ m or about 0.2 mm

(b) Time constant $= RC = (200 \times 10^6\ \Omega)(20 \times 10^{-12}\ \text{F})$

$= 4 \times 10^{-3}$ s or 4 ms

(i) The flexible plate will move in and out at 5000 Hz. If it moves with an amplitude of only 0.01 mm back and forward this is 5% of d, and so the capacitance, $C \propto 1/d$, of the microphone will vary at 5000 Hz.
(ii) At 5000 Hz changes in the sound wave are taking place in less than its period, $T = 1/f = 0.2$ ms, so the time constant of the circuit is very long, and the charge on the microphone can't change noticeably during a cycle of the sound wave.

(c) If C alters with the sound wave but Q is effectively constant then the p.d. across the microphone, $V = Q/C$, will vary at 5000 Hz.

As V (microphone) + V (200 MΩ resistor) = 100 V

the p.d. across the resistor will oscillate at 5000 Hz.

Example 4.23

(a) Using only three-input NOR gates design a system of which the output only goes high when all three inputs are high.
(b) What is the name for the logic gate you have designed?
(c) Give an advantage of being able to design logic circuits using only NOR gates. [5]

Solution 4.23

(a)

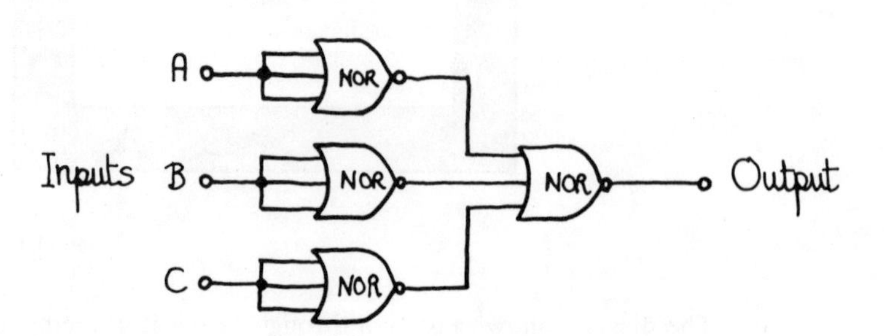

These circuits can be designed only by trial and error, but experience helps.

(b) It is a three-input AND gate.
(c) Chips are designed with several gates of the same type on them. It is more economical of space to design with just one type.

Also in the past a simple transistor switch could easily be used as a NOR gate.

Example 4.24

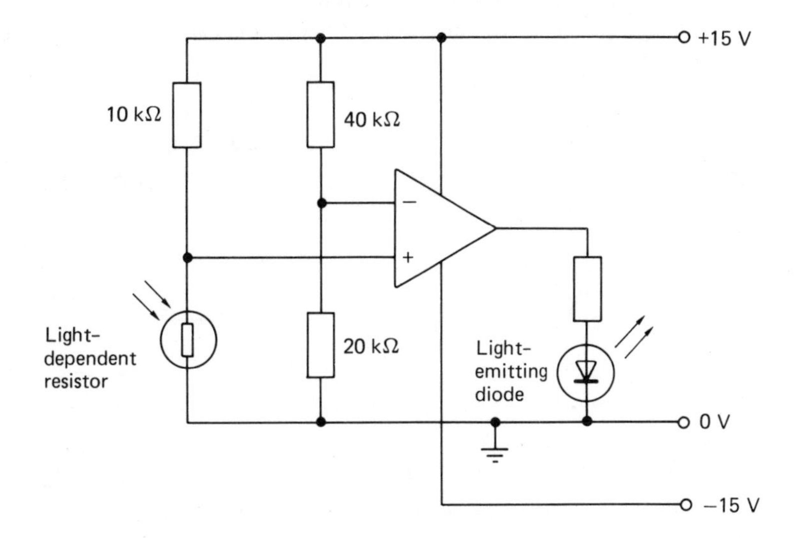

The circuit shown is designed to switch on a light-emitting diode (l.e.d.) when darkness falls. The circuit uses a light-dependent resistor (l.d.r.) which has a resistance greater than 1 MΩ in the dark, but less than 1 kΩ when illuminated.

(a) Calculate the potential at the inverting input.

(b) Calculate the approximate potential of the non-inverting input when the l.d.r. is
(i) in the dark
(ii) in the light.

(c) Why does the l.e.d. only light when darkness falls? [7]

Solution 4.24

(a) The inverting input (–) is connected to a potential divider.

$$\text{Potential of } - \text{ input} = \left(\frac{20\ \text{k}\Omega}{20\ \text{k}\Omega + 40\ \text{k}\Omega}\right)(15\ \text{V}) = 5\ \text{V}$$

(b) (i) $$\text{Potential of + input when dark} = \left(\frac{1\ \text{M}\Omega}{1\ \text{M}\Omega + 10\ \text{k}\Omega}\right)(15\ \text{V}) = 15\ \text{V}$$

(ii) $$\text{Potential of + input when lit} = \left(\frac{1\ \text{k}\Omega}{1\ \text{k}\Omega + 10\ \text{k}\Omega}\right)(15\ \text{V}) = 1.4\ \text{V}$$

10 kΩ ≪ 1 MΩ from the point of view of electronics calculations.

(c) 1.4 V < 5 V so when the l.d.r. is lit the + input has a lower potential than the – input, so the output of the op. amp. will be about –15 V. No current will flow in the l.e.d.

When the l.d.r. is in the dark the + input will be of a higher potential than the – input and so the output will be +15 V. A current will flow in the l.e.d. and it will light.

Note the resistor to protect the l.e.d. from damage.

Example 4.25

(a) Explain what is meant by negative feedback. [2]

(b) The diagram shows an operational amplifier circuit using feedback.

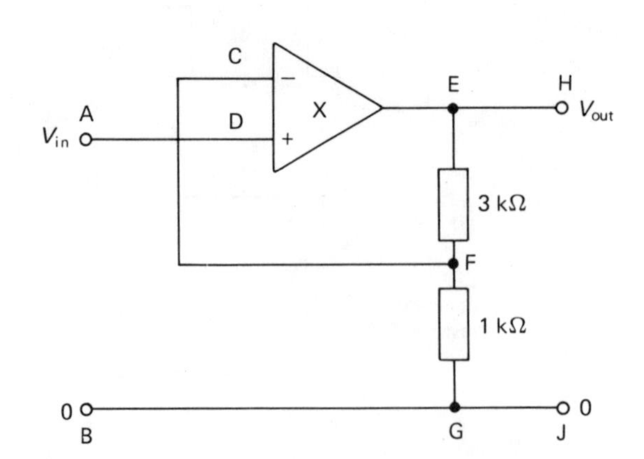

Give a reason for employing negative feedback in the circuit, and using the letters on the diagram, explain the negative feedback path. [2]

(c) The amplifier in the diagram has a gain of 80 000. What will V_{out} be when $V_{in} = 0.2$ V? [4]

Solution 4.25

(a) A fraction of the output signal of a device is fed back to the input, but in antiphase with the original input.

(b) The negative feedback improves the stability of the gain by making it dependent on the values of relatively stable resistors, and not on the very temperature-sensitive semiconductors in the operational amplifier.

The gain is greatly reduced in achieving this stability.

EFG is a potential divider across the output and the feedback path is then from F to C.

(c) The inherent gain of 80 000 is large so the gain of the amplifier with feedback is the reciprocal of the proportion of the signal that is fed back. Using the usual potential divider equation:

$$\text{Proportion fed back} = \frac{1\ \text{k}\Omega}{1\ \text{k}\Omega + 3\ \text{k}\Omega} = 1/4$$

$$V_{out} = 4 \times V_{in} = 4 \times 0.2\ \text{V} = 0.8\ \text{V}$$

Example 4.26

Read the following passage carefully and then answer the questions at the end.

The Digital Voltmeter (DVM)

A digital voltmeter gives a reading on a numerical display, whereas conventional moving coil meters give a reading which has to be interpreted from the position of a pointer on a scale. The digital meter thus gives an illusion of accuracy because the person reading it does not have to make any contributions to the estimation of the value or take precautions to avoid errors such as parallax.

The digital meter compares the voltage to be measured with an internal reference voltage which is a repetitive ramp waveform similar to the timebase of a c.r.o. The voltage to be measured is, after being amplified, fed in as one input to a comparator circuit as shown below. The reference voltage is fed into the other input. When the input, X, from the voltage

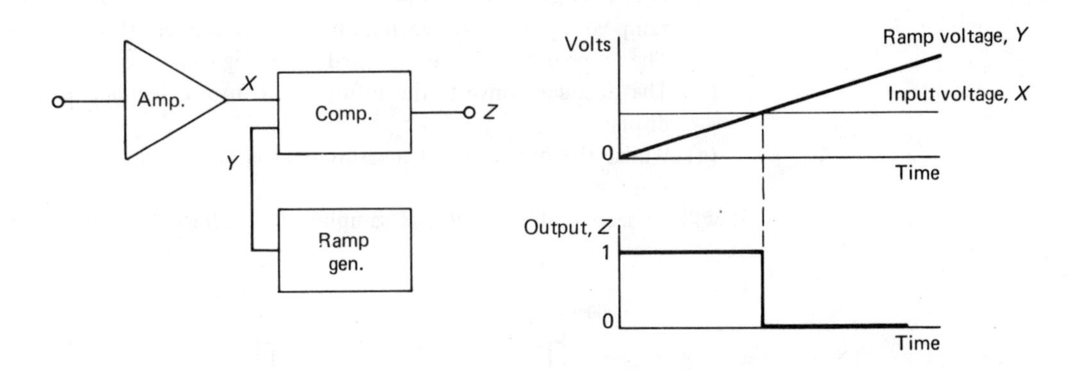

to be measured is greater than the reference voltage, Y, the output is 1. When Y exceeds X the output is 0. The output from the comparator, Z, is one input of the AND gate in the circuit below. The other input is a regular pulse train.

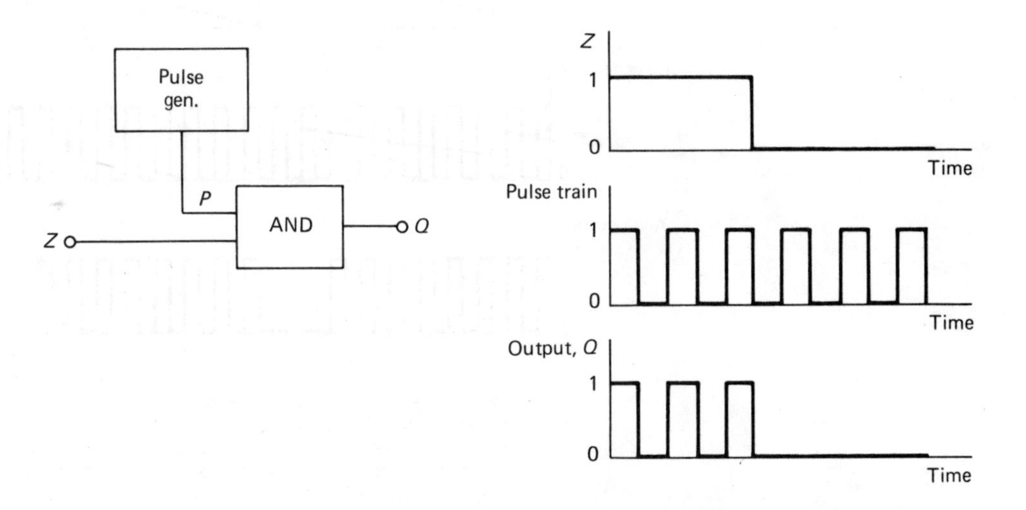

As long as the comparator output, Z, is 1 the pulse train is allowed through the AND gate. As long as $Z = 0$ the pulse train is cut off by the gate.

These pulses are fed into a counting circuit (register). A decoding circuit converts the information from binary to decimal form and it is then shown on a digital display. The complete system is thus as shown in block diagram form.

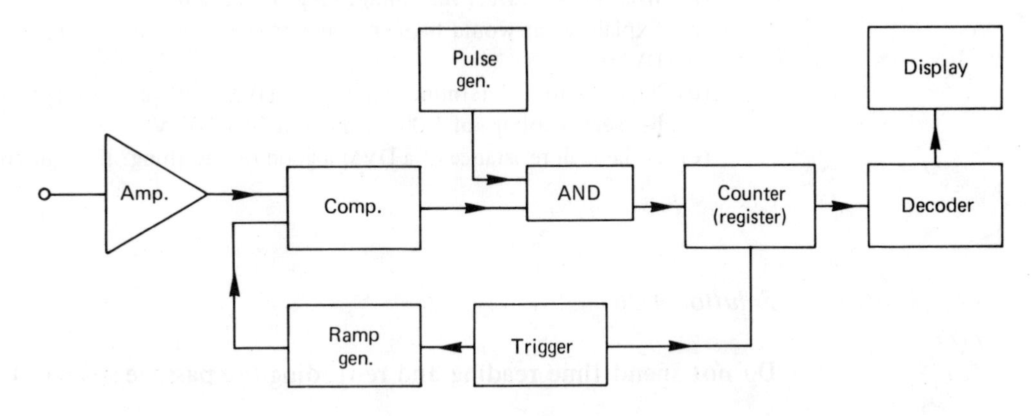

The cycle of operation is thus:

(a) A trigger starts the ramp and clears the register.

(b) The comparator gives a 1 output until the ramp voltage is equal to the voltage to be measured, X.

(c) When the ramp voltage reaches the value X the comparator output falls to 0.

(d) The AND gate admits a series of pulses, the number of which measures the time the ramp voltage takes to reach X. If the ramp is linear this time is proportional to X.

(e) The number of pulses is recorded in the register.

(f) The decoder converts the information into a decimal reading and passes it to the display.

(g) Whilst the operation of measurement is taking place the display holds the previous reading.

It will be seen that the DVM is sampling the voltage to be measured at regular intervals.

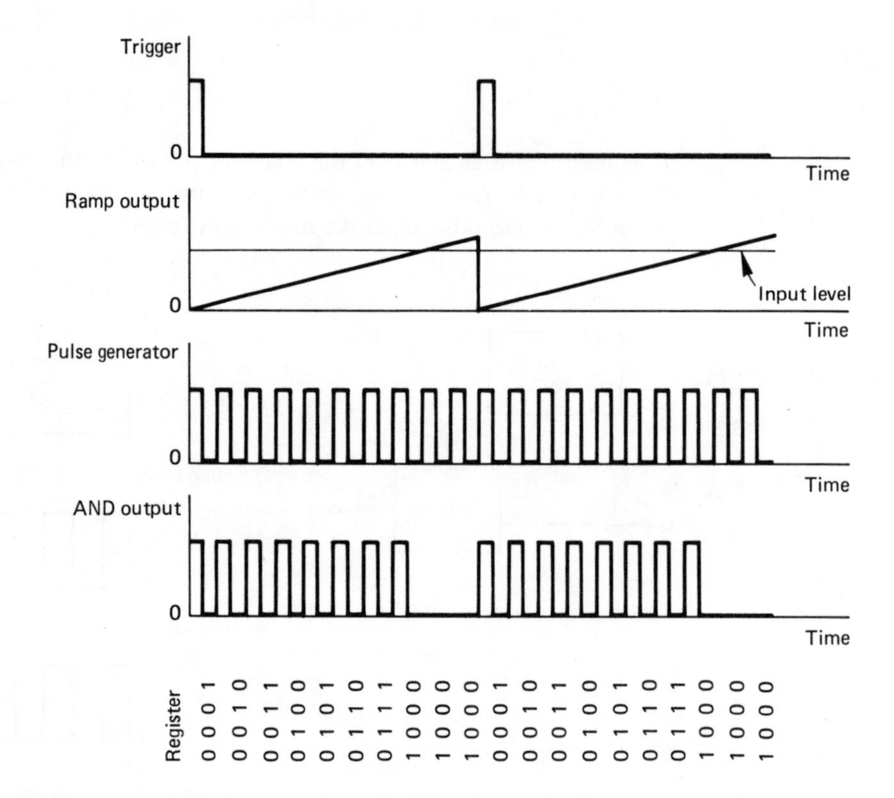

Digital voltmeters, because the input is into an electronic amplifier, can have a very high resistance – up to 10 MΩ in some cases. They can also measure very low voltages, of the order of millivolts, as the signal to be measured is amplified before being processed.

Questions

(a) Comment on the statement 'Instruments with digital displays are better because they are more accurate'. [3]

(b) What factors affect the voltage range of a DVM? [4]

(c) Explain what would happen if an alternating voltage were applied to the input of the DVM. [3]

(d) What factors determine whether a DVM with a four-digit display can distinguish between a voltage of 1.001 V and one of 1.002 V? [4]

(e) Is the high resistance of a DVM a good or bad thing? Explain your answer. [3]

[L]

Solution 4.26

Do not spend time reading and re-reading the passage so as to understand it fully.

(a) The accuracy of an instrument depends on the care with which it was initially calibrated and the stability of the components, resistors and capacitors, etc., within it. The reading of a digital meter is precise as there are no experimental errors caused by the person reading it but it may be very inaccurate, i.e. it may read + 13.46 V when the p.d. it is measuring is + 13.8 V.

Hard work for the marks. See the discussion of errors on page 197.

(b) The maximum p.d. that can be measured by a DVM is equal to the maximum of the ramp voltage which it generates. At the low p.d. end they can measure a few millivolts as the input is amplified.

(c) The DVM would sample the alternating voltage at regular intervals and the digital reading would change every time the register recorded the sampled voltage.

(d) To distinguish between 1.002 V and 1.001 V involves a pulse generator of high enough frequency to produce a pulse in the time it takes the ramp generator to rise 0.001 V.

You could go a lot further here, as is often the case in these questions. For example: a four-digit display measuring to 1 mV would have a maximum ramp voltage of 10 V, i.e. max p.d. is 9.999 V.

If the frequency of repetition of the ramp voltage is 100 Hz, then it must rise 10 V in 0.01 s or 1 mV in 10^{-6} s. If the pulse generator is to produce a pulse every 10^{-6} s, it must operate at 1 MHz. This is quite possible.

(e) A high resistance is a good thing. As a voltmeter measuring the p.d. across a resistor, a resistance of 10 MΩ means that very little current is taken by the DVM, even when connected across 100 kΩ.

Example 4.27

Draw a truth table for the logic system shown below, including the states at C, D, and E. [4]
(L)

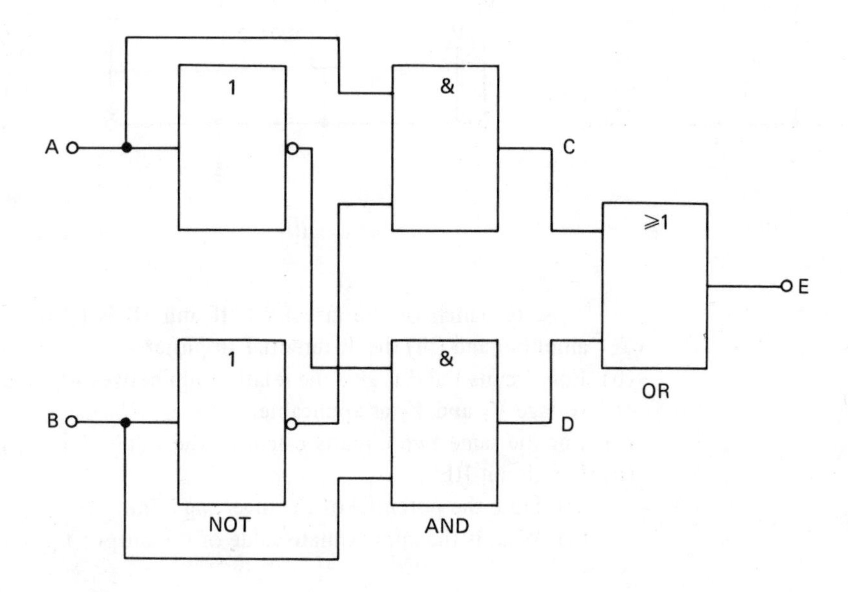

Solution 4.27

Points 1, 2, 3, and 4 are the inputs to the AND gates numbering down from the top.

A	B	1 (A)	2 ($\bar{B}$)	C	3 ($\bar{A}$)	4 (B)	D	E
0	0	0	1	0	1	0	0	0
1	0	1	1	1	0	0	0	1
0	1	0	0	0	1	1	1	1
1	1	1	0	0	0	1	0	0

$\bar{A}$ means the inverse of A and $\bar{B}$ the inverse of B.

Do not be afraid to label extra points. Basically such questions are simple, but you can make silly slips if you try to do too much in your head.

Example 4.28

(a) It is possible to use an op. amp. as a comparator, an inverting amplifier or as a differential amplifier. The diagrams show an op. amp. connected in three different ways. To simplify the circuit diagram the connections to the +9 V and –9 V supply have been omitted.

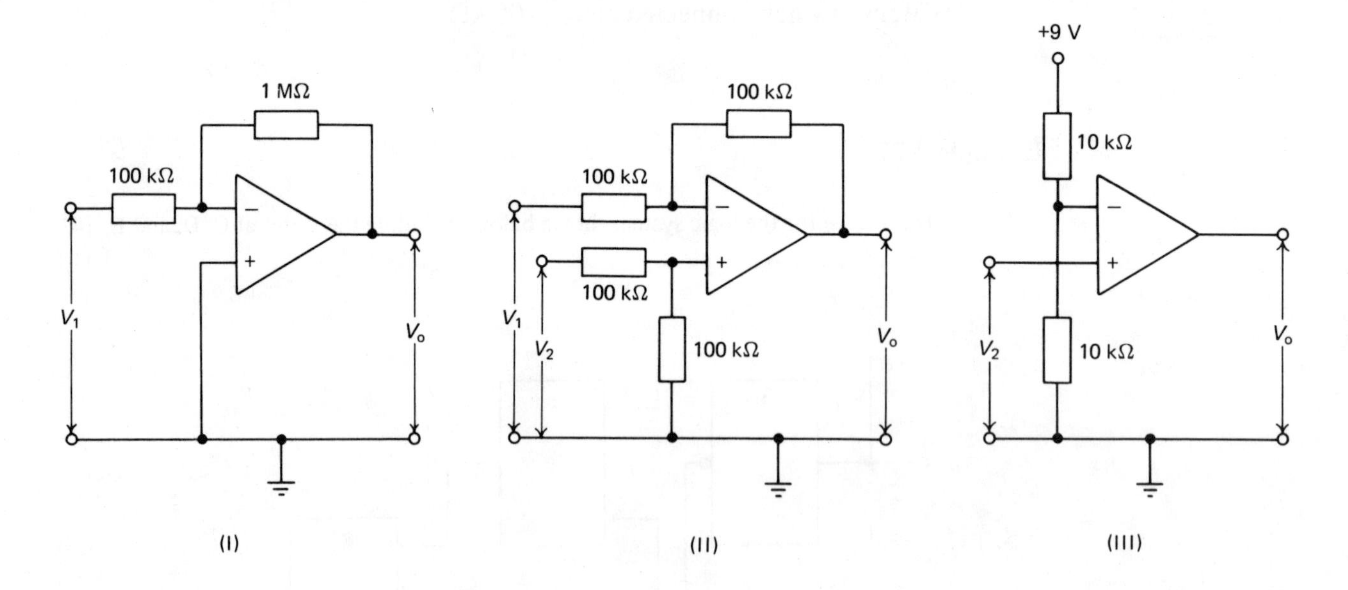

State which of the circuits I, II and III is (i) the comparator, (ii) the inverting amplifier and (iii) the differential amplifier.

(b) For circuits I and II give the relationship between the output voltage V_o and the input voltage V_1 and V_2 as applicable.

(c) For the same two circuits calculate the value of V_o when $V_1 = 0.5$ V and $V_2 = 2$ V.

(d) For circuit III

(i) State the potential of the inverting input.

(ii) What is the approximate value of the output V_o if $V_2 = 2$ V. [9]

(SEB)

Solution 4.28

(a) (i) III is a comparator.

The – input is fixed at 4.5 V and the + input has no feedback. If V_2 goes a little above 4.5 V the output will shoot up to its highest value owing to the huge inherent gain of the op. amp. If V_2 goes a little below 4.5 V the output will go to its lowest value.

(ii) I is an inverting amplifier.

Note the feedback resistors and that the – input is used while the + is grounded.

(iii) II is a differential amplifier.

Both inputs are used so the output is proportional to the difference in potential between them. The gain for each input is reduced to 1 and so it could be used as a subtractor.

(b) $$V_o = \frac{-R_f V_1}{R_1} = \frac{-(1000 \text{ k}\Omega) V_1}{(100 \text{ k}\Omega)} = -10 V_1$$

$$V_o = \frac{(100 \text{ k}\Omega) V_2}{(100 \text{ k}\Omega)} - \frac{(100 \text{ k}\Omega) V_1}{(100 \text{ k}\Omega)} = V_2 - V_1$$

(c) $V_o = -10 V_1 = -10 (0.5 \text{ V}) = -5 \text{ V}$

$V_o = V_2 - V_1 = 2 \text{ V} - 0.5 \text{ V} = 1.5 \text{ V}$

(d) (i) 4.5 V

The two resistors of the potential divider are the same and the p.d. across the whole of it is 9 V.

(ii) $V_2 < V_1$ when V_2 is 2 V and so the output will be about –9 V.

Example 4.29

Give one example of positive feedback experienced in electrical circuits. [3]

Solution 4.29

When the current in a transistor rises, the energy transformed in it rises also. This may result in the temperature of the transistor going up and hence in its resistance falling, so that the current may rise further. If this process of positive feedback continues, the transistor junction may melt. This effect is called thermal runaway.

See also question 4.19 on page 116 for other everyday examples of feedback.

Example 4.30

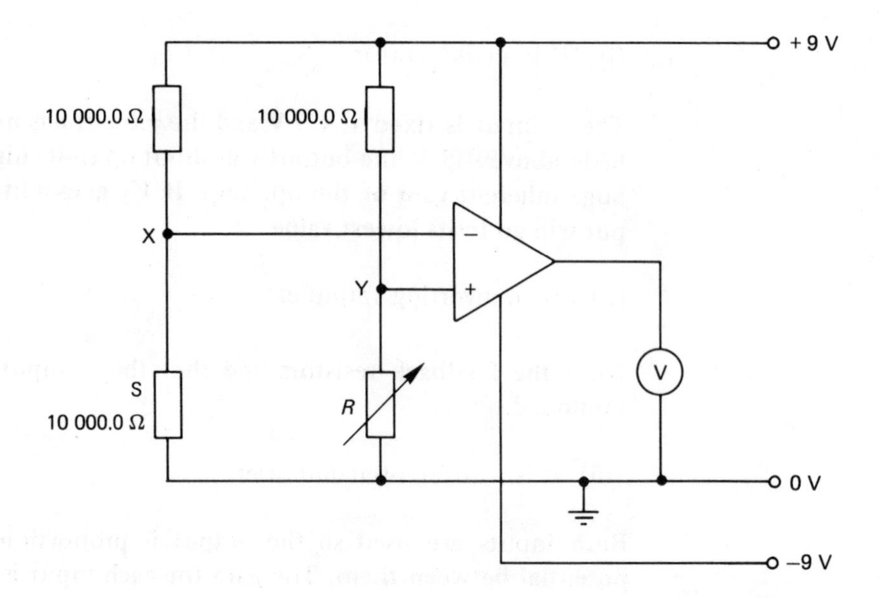

The circuit shows 4 very high quality resistors arranged as 2 potential dividers. R is variable from 0 to 19 999.9 Ω. The potentials of X and Y, the midpoints of the potential dividers, are compared using an operational amplifier. The inherent voltage gain of the amplifier, A = 100 000. You may assume that it has an infinitely high input resistance, that is its inputs draw no current.

(a) (i) Calculate the potential of point X.

(ii) What potential difference must there be between X and Y if the voltmeter is just to read +9 V?

(iii) Which of X and Y is at the higher potential? [3]

(b) The voltmeter just reads +9 V when R is set at 10 000.4 Ω. Sketch a graph of how the voltmeter reading will vary (y-axis) as R is varied in the range 9 999.0 Ω to 10 001.0 Ω (x-axis). [2]

(c) Resistor S is replaced by a strain gauge. This is a zigzag arrangement of very thin insulated wire which is firmly glued to a metal beam. As the beam is bent the wire is stretched or compressed and the resistance of the wire changes very slightly. The strain gauge is known to have a resistance of 10 kΩ, that is between 9.5 kΩ and 10.5 kΩ. It is also known that the resistance of the gauge changes by 0.4 Ω when a 1 kg mass is hung on one end of the beam.

(i) Explain how you could use the circuit and beam to measure masses in the range 0 to 1 kg.

(ii) Instead of connecting points X and Y to the operational amplifier a galvanometer is connected between them. It has a resistance of 50 Ω and is graduated in divisions of 0.1 mA. Would this also indicate the strain due to a 1 kg mass hanging on the beam? Explain your answer using calculations as necessary. [5]

Solution 4.30

(a) (i) The two resistors in the potential divider are the same so X must have a potential halfway between 0 and 9 V, that is 4.5 V.

(ii) (Inherent gain) (p.d. between inputs) = +9 V

$$\text{p.d.} = \frac{9\text{ V}}{100\,000} = 9 \times 10^{-5}\text{ V} = 90\ \mu\text{V}$$

(iii) Y

The non-inverting input must have the higher potential.

(b)

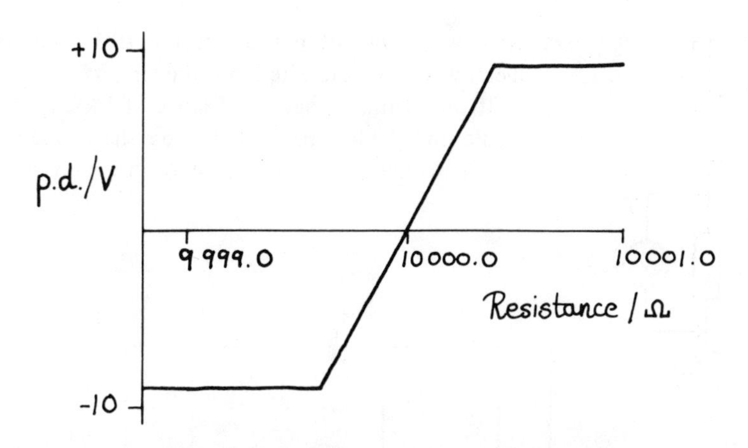

(c) (i) R must be adjusted until it is exactly the same resistance as S. The voltmeter reads 0 when this is the case.

Now if masses are added the voltmeter reading will rise until it reaches +9 V when a 1 kg mass is straining the beam. To find an unknown mass one uses the formula

$$\text{Unknown mass} = \left(\frac{\text{voltmeter reading for unknown mass}}{9\text{ V}}\right)(1\text{ kg})$$

(ii) The galvanometer will only register if a p.d. greater than (0.0001 A) (50 Ω) = 0.005 V is applied across its terminals.

This is considerably higher than the 0.000 09 V caused by the strain gauge. The galvanometer will not indicate the stress due to a 1 kg mass.

4.9 Questions

Question 4.1

In the circuit shown below, the 6 V battery has negligible internal resistance.

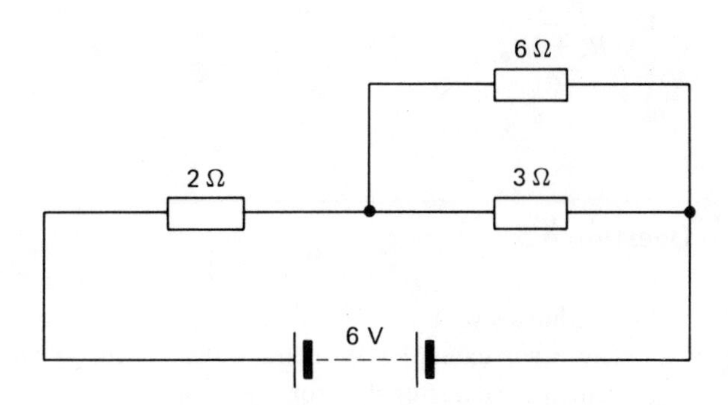

What is the current in the 6 Ω resistor?

A 0.5 A **B** 1.0 A **C** 1.5 A **D** 2.0 A **E** 2.5 A (OLE)

Question 4.2

Draw a labelled diagram of a potentiometer arrangement for obtaining a variable 0 V–12 V d.c. supply from a fixed 12 V d.c. source. [3]

(L)

Question 4.3

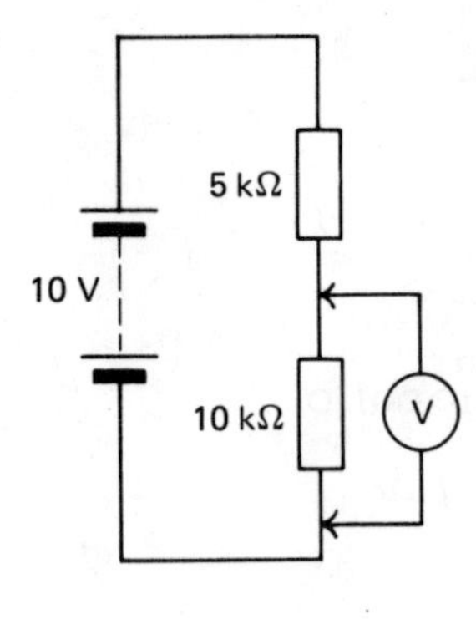

(a) What would be the p.d. across the 10 kΩ resistor in the circuit shown if the voltmeter were not connected into the circuit?

If the voltmeter has a resistance of 1 kΩ per volt and is set on its 10 V scale what would it read when connected across the 10 kΩ resistor?

What conclusion do you draw from these values?

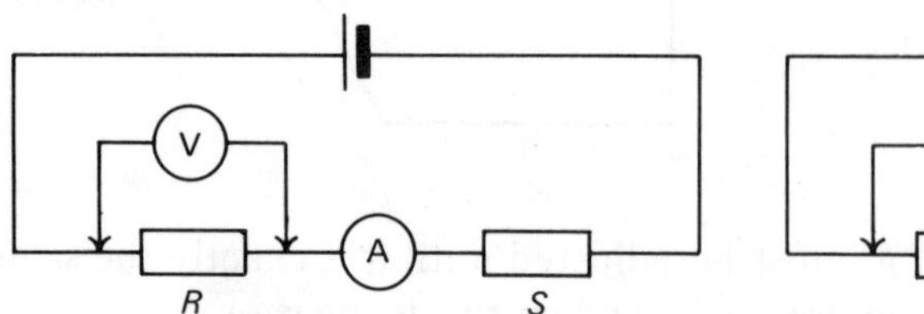

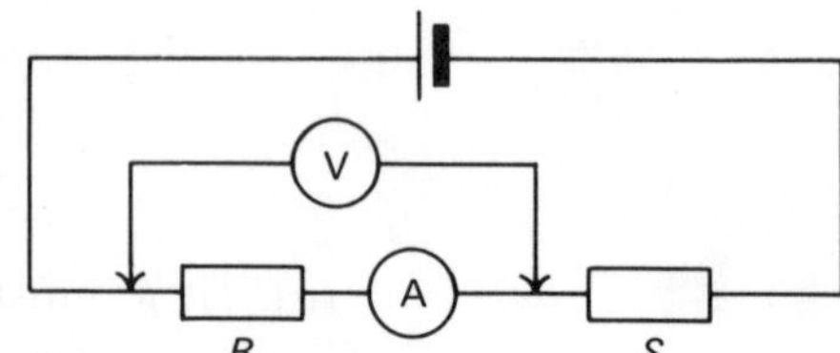

(b) The diagrams above show two ways of measuring the p.d. across and current through a resistor. State, giving your reasons, which arrangement is to be preferred if

(i) $R = 2\ \Omega$ and $S = 4\ \Omega$,

(ii) $R = 500\ \text{k}\Omega$ and $S = 1\ \text{M}\Omega$. [12]

(L)

Question 4.4

A length of high-voltage overhead electrical cable consists of a steel core surrounded by six aluminium conductors, as shown in the sketch of the cross-section of the cable. If the resistance of the steel core is R_s and that of the aluminium conductors is R_a, the resistance of the composite cable is

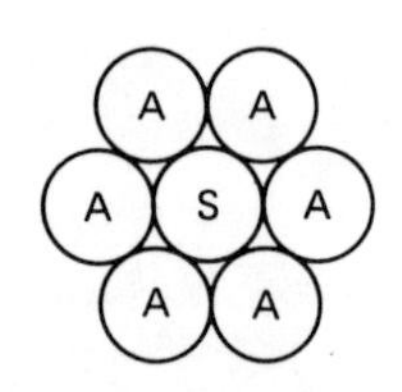

A = aluminium
S = steel

A $R_s + 6R_a$

B $\dfrac{1}{R_s} + \dfrac{1}{6R_a}$

C $\dfrac{1}{R_s} + \dfrac{1}{R_a}$

D $\dfrac{R_sR_a}{R_a + 6R_s}$

E $\dfrac{R_sR_a}{R_s + 6R_a}$

(NISEC)

Question 4.5

State Ohm's law.

Give two examples of conductors which do *not* obey Ohm's law, and sketch typical current/p.d. characteristics for each. [6]

Question 4.6

Two resistors P and Q have the same size and shape and have similar surfaces. When connected in series with a low voltage power supply of negligible resistance, P becomes much hotter than Q; when connected in parallel with the same power supply, Q becomes the hotter.

Give a full explanation for this and deduce which resistor has the bigger resistance. [6]

Question 4.7

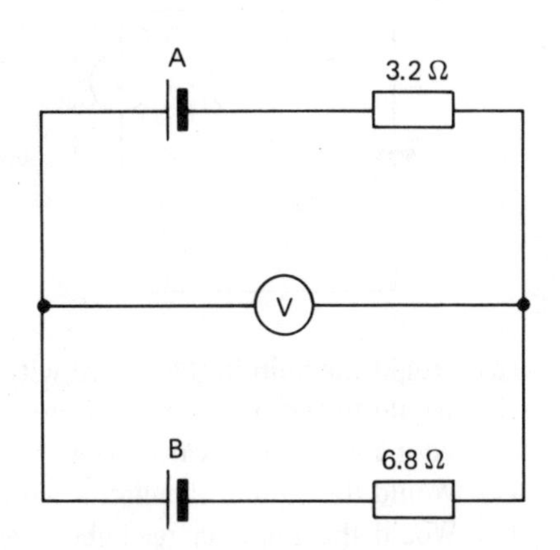

A is a cell with an e.m.f. of 2.2 V and an internal resistance of 0.8 Ω.
B is a cell with an e.m.f. of 1.5 V and an internal resistance of 1.2 Ω.
The voltmeter shown is a digital one and its resistance may be taken to be infinite. Calculate the reading on the voltmeter. [6]

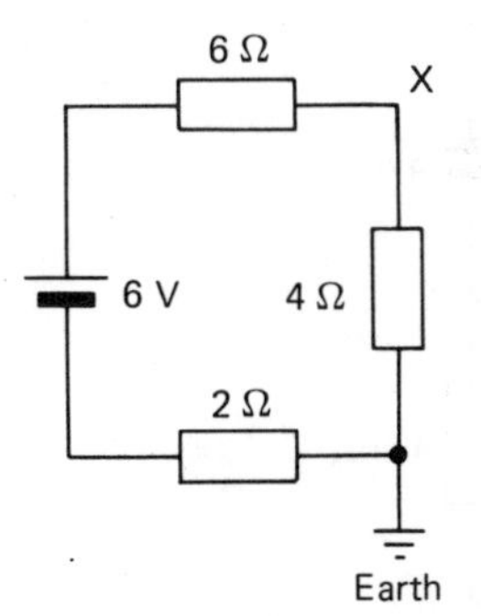

Question 4.8

The circuit shown consists of three resistors with a 6 V battery of negligible internal resistance supplying the e.m.f. The potential, in V, at the point X is

A +6 **B** +3 **C** +2 **D** –2 **E** –4 (NISEC)

Question 4.9

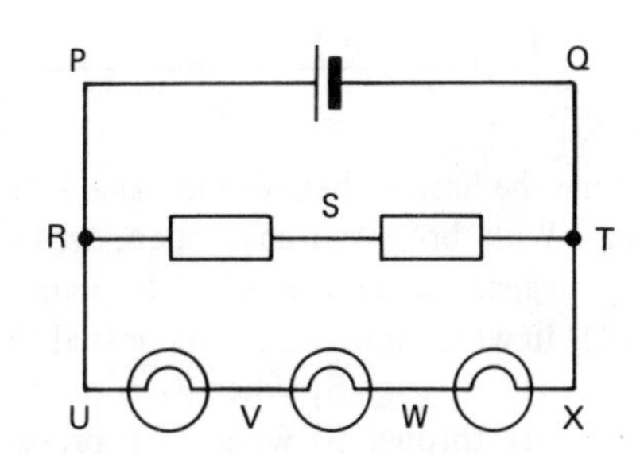

In the circuit shown, the resistors are identical, as are the lamps. When a voltmeter is connected between R and T a reading of 6 V is obtained. In the questions below you are asked to deduce the reading of the voltmeter when it is connected between various other points in the circuit. The voltmeter reading is to be selected from the list (lettered **A** to **E**) below.

A 0 V **B** 1 V **C** 3 V **D** 4 V **E** 6 V

(a) What is the reading of the voltmeter when it is connected between U and W?
(b) What is the reading of the voltmeter when it is connected between S and T?
(c) What is the reading of the voltmeter when it is connected between V and S?
(d) One of the lamps is removed from its socket. Assuming the cell has zero internal resistance, what is the reading of the voltmeter when it is connected between U and X? (NISEC)

Question 4.10

The circuit diagrams overleaf show a voltmeter and an ammeter *wrongly* connected. The cell has an e.m.f. of 2 V and internal resistance 0.45 Ω. The voltmeter has a range between 0 and 2 V and internal resistance of 20 000 Ω. The ammeter has a range between 0 and 4 A and a resistance of 0.05 Ω. The bulb is rated 2 V, 0.2 A.

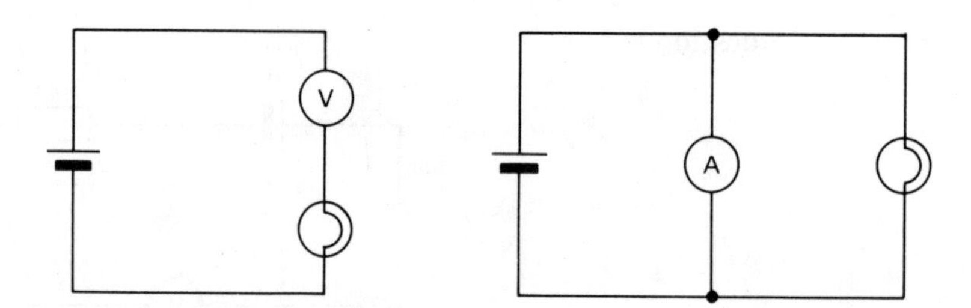

(a) Would the bulb in the circuit with the voltmeter be lit? Explain your answer. [1]
(b) Would the voltmeter read about zero, roughly half scale or close to full scale? Explain your answer using calculations if you wish. [2]
(c) Would the bulb in the circuit with the ammeter be lit? Explain your answer. [2]
(d) Would the ammeter read about zero, roughly half scale or close to full scale? Explain your answer using calculations if you wish. [2]

Question 4.11

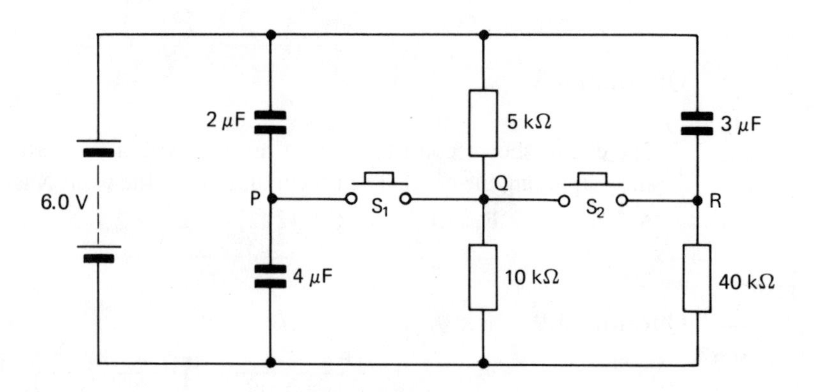

Assume the battery has zero internal resistance.

(a) With both switches open, write down the potential difference across each of the capacitors and resistors. Explain your thinking.
(b) In which direction is the initial flow of charge
  (i) through S_1 when S_1 is pressed?
  (ii) through S_2 when S_2 is pressed?
(c) Calculate the charge stored on each of the three capacitors when both S_1 and S_2 are pressed. [12]

Question 4.12

(a) Define *capacitance* of a capacitor and state the unit in which it is measured. Derive an expression for the effective capacitance of two capacitors of capacitances C_1 and C_2 arranged in series. [5]
(b) Describe, with the aid of a diagram, the structure of a practical form of capacitor. What is meant by its 'working voltage'?
  Explain on what feature of the capacitor which you have described the working voltage chiefly depends. [8]
(c) Explain qualitatively why an alternating current flowing in a circuit containing a capacitor and resistor in series may be varied by changing (i) the capacitance of the capacitor and (ii) the frequency of the alternating supply. [4]
(d) A capacitor of capacitance C is being charged through a resistor of resistance R. Show that the quantity RC has the unit of time. What is this quantity called? What information does it give you concerning the charging of the capacitor? [4]

(AEB 1983)

Question 4.13

Three 1.0 μF capacitors are (a) connected in series to a 2.0 V battery, (b) connected in parallel with each other and a 2.0 V battery. Calculate the charge on each of the capacitors in each of the cases (a) and (b).

Account, without calculation, for the difference in energy stored in each capacitor in cases (a) and (b). [6]

(L)

Question 4.14

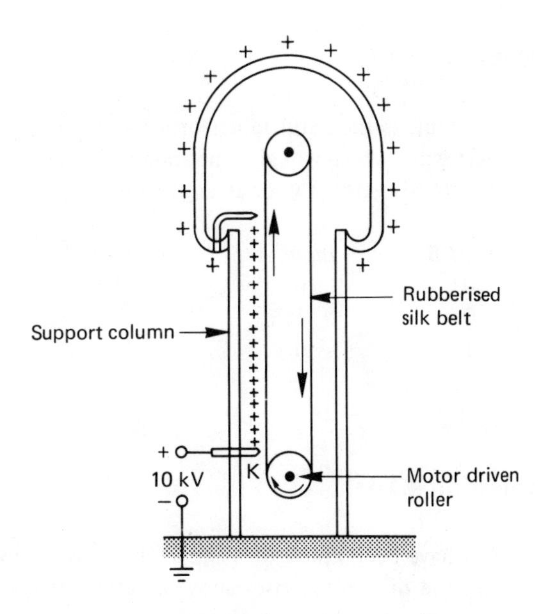

The diagram shows a Van de Graaff generator in which charge sprayed on the belt from the comb K is carried to the upper dome, thus raising it to a high potential.

(a) What type of material should the support column be made of? If the charge on the belt is 2 μC m^{-2} and is only on one side of the belt which is 5 cm wide, what quantity of charge reaches the upper dome in 8.0 s when the belt is moving at a speed of 0.6 m s^{-1}? [4]

(b) A spark discharge then transfers all this charge to a nearby earthed conductor in 30 μs.

(i) What is the average current during the sparking?

(ii) When this charging and discharging of the dome continues regularly, it is noticed that the motor driving the lower roller tends to slow a little before and speed up immediately after each discharge. Suggest a reason for this behaviour. [4]

Question 4.15

A capacitor is labelled 'Electrolytic, 3000 μF, 25 V'.

(a) How much charge does the capacitor store when fully charged?

(b) How much energy does the capacitor store when fully charged?

(c) What might happen if a potential difference much greater than 25 V where applied across the capacitor? [7]

Question 4.16

(a) Draw a labelled cross-sectional diagram of an electrolytic capacitor. Explain why it is important to connect this device with the terminals having the correct polarity.

(b) What is the capacitance of two capacitors of 2 μF and 4 μF connected (i) in series, (ii) in parallel?

(c) Draw sketch graphs of

(i) the charge stored on the capacitor,

(ii) the current through the resistor

against time, starting at the moment when the switch in the circuit shown is closed.

Mark approximate values on the axes of your graphs. [15]

(L)

5 V

10 μF

100 kΩ

Question 4.17

The table is the truth table for a two-input digital system. Sketch a simple arrangement of logic gates which will give the stated outputs, using any of NOT (INVERT), OR, NOR, AND and NAND gates in your arrangement.

Input B	*Input A*	*Output C*
0	0	0
0	1	1
1	0	1
1	1	0

[7]

Question 4.18

You have been asked to make a simple system which will cause a bell to ring when the temperature of a room rises above a certain temperature in the range 30°C to 40°C. You are given an operational amplifier, a power amplifier, a relay, an electric bell and a thermistor with the following characteristic:

temperature/°C	30	35	40
resistance/Ω	600	530	480

(a) With the aid of a block diagram describe how you would arrange your system.

(b) Use a circuit diagram to explain how you would allow the temperature at which the alarm rings to be altered.

Give suitable values for the components used and explain your choice. [10]

Question 4.19

The following are examples of feedback taken from everyday life. State in each case whether the feedback is positive or negative and briefly explain your choice.

(a) As a room at a party gets noisier people talk louder to make themselves heard.

(b) As vegetables go out of season their price rises. [4]

4.10 Answers to Questions

4.1 It is not necessary to find the current from the cell (which is 1.5 A) in order to get the answer.

Answer A

4.2 See page 77.

4.3 (a) Draw the circuit both (i) without the voltmeter, and (ii) with it connected. See worked examples 4.8 and 4.10 for potential dividing calculations.
Answers 6.7 V, 5.0 V

(b) Give values to the resistance of both voltmeter, say 10 kΩ, and the ammeter, say 1 Ω.

(i) Draw *both* circuits adding resistance values for R, S, A and V. Discuss the fraction of current in V in both circuits and fraction of the potential dropped across A.

(ii) Draw both circuits again

Yes, *four* circuit diagrams in answer to part (b).

adding the resistance values. Discuss the current in V and the p.d. across A as in (i).

4.4 The six aluminium conductors in parallel each of resistance R_a have a net resistance of $R_a/6$. This, in parallel with the steel core of resistance R_s, gives a total resistance R, where

$$\frac{1}{R} = \frac{1}{R_s} + \frac{1}{R_a/6} = \frac{1}{R_s} + \frac{6}{R_a} = \frac{R_a + 6R_s}{R_a R_s}$$

$$\Rightarrow \quad R = \frac{R_a R_s}{R_a + 6R_s}$$

Answer **D**

4.5 See the introduction to this chapter; in particular section 4.4.

4.6 Draw the two circuits, labelling the resistors R_P and R_Q. For the series circuit let the *common current* be I and write down expressions for the power transformed in P and Q. For the parallel circuit let the *common p.d.* be V and write down the new power expressions.

P is hotter in series, Q is hotter in parallel. This means the power transformed in P is the greater when they are in series, etc. You should have quite enough to answer this difficult but important question.

4.7 Draw a circuit diagram which includes the internal resistances of the cells, but excludes the voltmeter which does not affect the circuit. Transfer the data from the question to your diagram. Then calculate

(i) the total e.m.f., being careful about direction,

(ii) the total resistance around the circuit, and hence

(iii) the current in the circuit.

The p.d.'s across each resistor and across the internal resistance of each cell can now be found and the p.d. across the voltmeter calculated.
Answer 2.0 V

4.8 Answer **C**
It is not necessary to calculate the current but it is very easy to do so in this case and enables you to find the p.d. across the 4 Ω resistor.

The earth connection makes no difference to the charge flow round the circuit – it doesn't 'leak out'.

Try writing down the potentials if the point X was earthed (i) *instead of* and (ii) *as well as* the point shown.

4.9 The answers are **D**, **C**, **B**, **E** to the four parts. This is an important question. Try writing down the p.d.'s (using a notation such as V_{RS}, etc.) between each pair of points given.

4.10 (a) No. Consider the three resistances in the circuit and get a rough idea of how big the current is.
(b) Full scale. Think of the bulb and voltmeter as a potential divider and remember that even voltmeters with a low resistance really do indicate the potential difference between their terminals.
(c) No. You should consider why 'short circuiting' the bulb stops it lighting. Remember the 0.45 Ω.
(d) Nearly full scale. The bulb can be ignored when considering the total resistance of the circuit.

4.11 (a) 2 μF – 4 V, 4 μF – 2 V, think of charge.
5 kΩ – 2 V, 10 kΩ – 4 V, think of current.
3 μF – 6 V, 40 kΩ – 0 V.
(b) (i) From Q to P. (ii) From Q to R.
(c) 2 μF – 5 μC, 4 μF – 15 μC, 3 μF – 7 μC

4.12 (a) Remember when in series the charge on each capacitor is the same. Think why this must be so. Look up the proof in a textbook if necessary.
(b) The use of a coloured pen on your diagram might help here. Check with your textbook.
(c) Think carefully about the maximum value of the charge flowing on and off the capacitor, and how often it must flow on and off each second. See page 97.
(d) This is asking you to show that Ω times F is equivalent to s. See the table on page 73 and reduce the units as necessary to their basic components. See also worked example 1.5.

4.13 Draw simple circuit diagrams to help clarify your ideas about series and parallel.
Answers
(a) 6.7×10^{-7} F
(b) 2.0×10^{-6} F
Think about how much charge is stored on each capacitor and the average p.d. during the charging process. Then remember the volt is a joule per coulomb.

4.14 (a) 0.48 μC
(b) 16 mA; consider forces between like charges.

4.15 (a) (b) See worked examples 4.16 and 4.17.
Answers 7.5×10^{-2} C, 0.94 J
(c) This is about the insulator, which is a very thin layer of oxide on one side of the aluminium electrode.

4.16 (a) The insulator which separates the 'plates' of metal and liquid is the thin oxide layer – beware.
(b) (i) 1.3 μF (ii) 6 μF

(c) For the current see worked example 4.22 and use the time constant RC, as the time for the current to fall to 1/e (about a third) of its initial value, to mark the time axis.

The initial current is 50 μA and the final charge stored is 50 μC – you should calculate each value.

The graph of charge stored mirrors that of the current, i.e. it rises quickly at first and approaches its maximum value exponentially – see page 78.

4.17 The output is that of an exclusive OR (EOR) used in half adders.

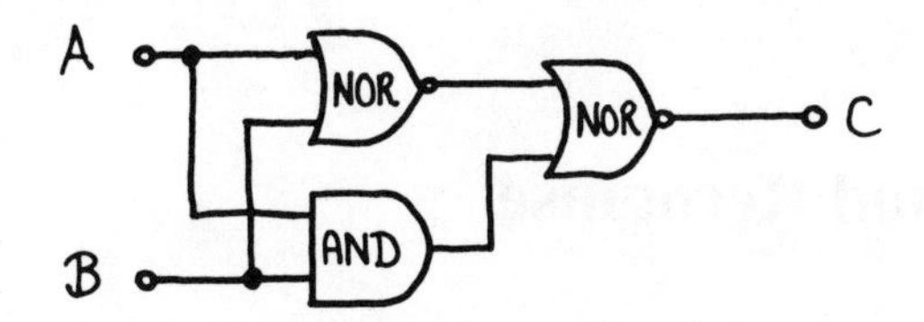

4.18 Both inverting and non-inverting inputs are connected to potential dividers and the op. amp. acts as a comparator. A variable resistor in one of the potential dividers will make the system adjustable. The relay and power amplifier are needed because the op. amp. cannot supply the necessary current to ring the bell.

See worked example 4.30.

4.19 (a) is positive and (b) is negative.

5 Electrostatics and Electromagnetism

5.1 You Should Recognise

Quantity	*Symbol*	*Unit*	*Comment*
electric field strength	E	newton per coulomb, $N\,C^{-1}$	or volt per metre, $V\,m^{-1}$
permittivity of vacuum	ϵ_0	farad per metre, $F\,m^{-1}$	a constant = $8.85 \times 10^{-12}\ F\,m^{-1}$
electric charge density	σ	coulomb per metre squared, $C\,m^{-2}$	refers to a surface charge
electric potential	V	volt, V	$1\ V = 1\ J\,C^{-1}$
magnetic flux density	B	tesla, T	$1\ T = 1\ N\,m^{-1}\,A^{-1}$ Earth's field $\approx$ 50 μT
permeability of vacuum	μ_0	henry per metre, $H\,m^{-1}$	a constant = $4\pi \times 10^{-7}\ H\,m^{-1}$
magnetic flux	Φ	weber, Wb	$1\ Wb = 1\ T\,m^2 = 1\ V\,s$
number of turns of wire	N	(no unit)	
number of turns per unit length	n	per metre, m^{-1}	e.g. of a solenoid
relative permeability	μ_r	(no unit)	over 1000 for soft ferromagnetic materials
self-inductance	L	henry, H	$1\ H = 1\ V\,s\,A^{-1}$
mutual inductance	M	henry, H	
frequency	f	hertz, Hz	mains frequency = 50 Hz
r.m.s. value, e.g. for alternating current	I_{rms}	ampere, A	r.m.s. means root mean squared
peak value, e.g. for alternating p.d.	V_0	volt, V	

Not listed are quantities, such as mass, length and time, which occur in all chapters and which are given in the list on page 7.

5.2 You Should be Able to Use

- Electric fields:

  Coulomb's law of force for point charges $F = \dfrac{Q_1 Q_2}{4\pi\epsilon_0 r^2}$

  electric field strength $E = \dfrac{F}{Q}$

  for a uniform field $E = \dfrac{V}{d}$

for a point charge $E = \dfrac{Q}{4\pi\epsilon_0 r^2}$

electric potential difference in an electric field $\Delta V = E\Delta x$

- Magnetic fields:
  force on a current $F = BIl$ (B perpendicular to l)
  force on a moving charge $F = BQv$ (B perpendicular to v)
  magnetic flux density
  - inside a long solenoid $B = \mu_0 nI$
  - near a long straight wire $B = \dfrac{\mu_0 I}{2\pi r}$
  - at the centre of a flat coil $B = \dfrac{\mu_0 NI}{r}$
- Electromagnetic induction:
  magnetic flux $\Phi = BA$ (B perpendicular to A)
  magnetic flux linking a coil = $N\Phi$

induced e.m.f. = rate of change of flux linking a circuit $V_{ind} = -N \dfrac{d\Phi}{dt}$

Faraday's law is $V_{ind} \propto \dfrac{d\Phi}{dt}$ the minus sign is Lenz's law

induced e.m.f. across a moving conductor $V_{ind} = Blv$

self-inductance induced e.m.f. = $-L \dfrac{dI}{dt}$

for a transformer $\dfrac{V_s}{V_p} \approx \dfrac{N_s}{N_p}$ $I_s V_s \leqslant I_p V_p$

mutual inductance induced e.m.f. = $-M \dfrac{dI}{dt}$

- Sinusoidal alternating currents:
  instantaneous, peak and r.m.s. values

$I = I_0 \sin 2\pi ft$ $V = V_0 \sin 2\pi ft$

$I_{rms} = \dfrac{I_0}{\sqrt{2}}$ $V_{rms} = \dfrac{V_0}{\sqrt{2}}$

power in pure resistors $P = I^2_{rms}R = \dfrac{V^2_{rms}}{R} = \frac{1}{2} I_0 V_0$

reactance of a capacitor $X_C = \dfrac{1}{2\pi fC}$

reactance of an inductor $X_L = 2\pi fL$

resonant frequency of an LRC circuit $f_r = \dfrac{1}{2\pi}\sqrt{\dfrac{1}{LC}}$

5.3 Electrostatics

Bodies can be charged by friction (contact). As charge is conserved, a positive charge is always associated with an equal negative, e.g. induced, charge.

Like charges repel, unlike charges attract.

The electric fields produced by charged bodies are described using lines of force. Perpendicular to these lines are equipotential surfaces – see Fig. 5.1. A uniform layer of charge produces a uniform electric field of strength $E = \sigma/\epsilon_0$. A point charge produces a radial field, $E \propto 1/r^2$ (see Fig. 5.2) and a uniformly charged

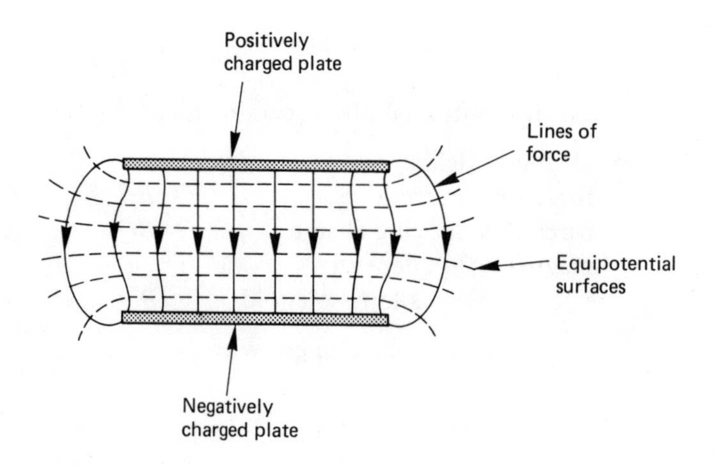

Figure 5.1

conducting sphere produces (i) zero field inside, and (ii) an inverse square radial field outside.

In uniform fields, differences of electric potential energy, e.p.e., are given by $QE\Delta x$; the difference in e.p.e. per unit charge is called the (electric) potential difference or p.d., ΔV. The p.d. between any two points on a charged conducting body is zero.

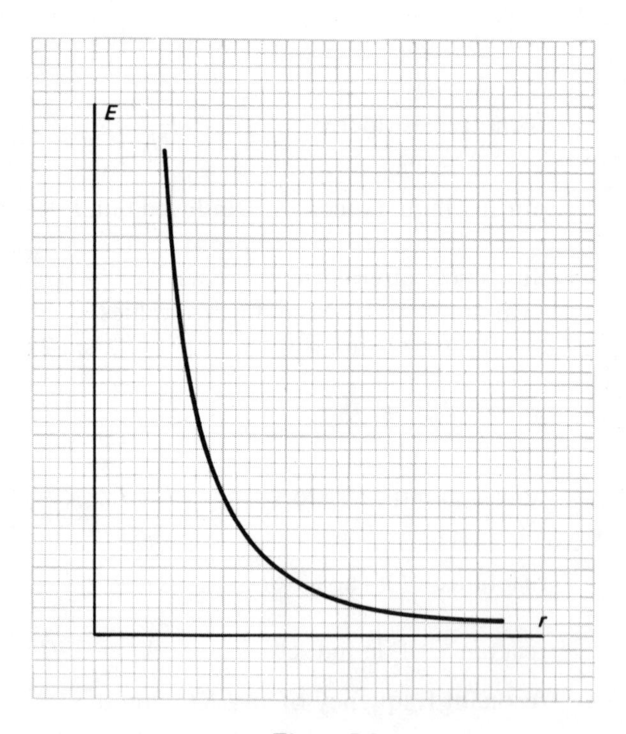

Figure 5.2

5.4 Electromagnetism

(a) Magnetic Fields

Magnetic field lines or lines of force form continuous closed loops. A rectangular coil with its axis perpendicular to a uniform magnetic field experiences a couple when there is a current in the coil. The design of moving coil meters and electric motors uses this couple.

The magnetic field of a current I_1 in a wire produces a force on a length l of a parallel wire in which there is a current I_2. If the wires are a distance r apart in vacuum, the force is

$$F = \frac{\mu_0 I_1}{2\pi r} I_2 l$$

This forms the basis of the definition of the ampere and fixes μ_0 at the value $4\pi \times 10^{-7}$ N A^{-2}.

A flow of charged particles moving perpendicular to a magnetic field in a conducting wire experience a force. This can lead to a p.d. across the conductor and is called the Hall effect.

(b) Electromagnetic Induction

Induced e.m.f.'s produce induced currents which are always in such a direction as to oppose the inducing effect.

Induced e.m.f.'s, V_{ind}, can usefully be split into two sorts:

(i) When there is a varying magnetic field through a fixed circuit or coil, $V_{ind} \propto \frac{dB}{dt}$. Search coils use this effect to compare magnetic field strengths.

(ii) When a coil or circuit moves in a uniform magnetic field, $V_{ind} \propto \frac{dA}{dt}$. In d.c. motors this results in an increasing induced e.m.f. as the speed of rotation of the motor rises.

Eddy currents can result from either (i) or (ii), giving rise to heating effects.

(c) Inductance

An inductor produces an induced e.m.f. when the current in it changes. When, in Fig. 5.3, the switch is closed the growth of current in the circuit is as shown in the

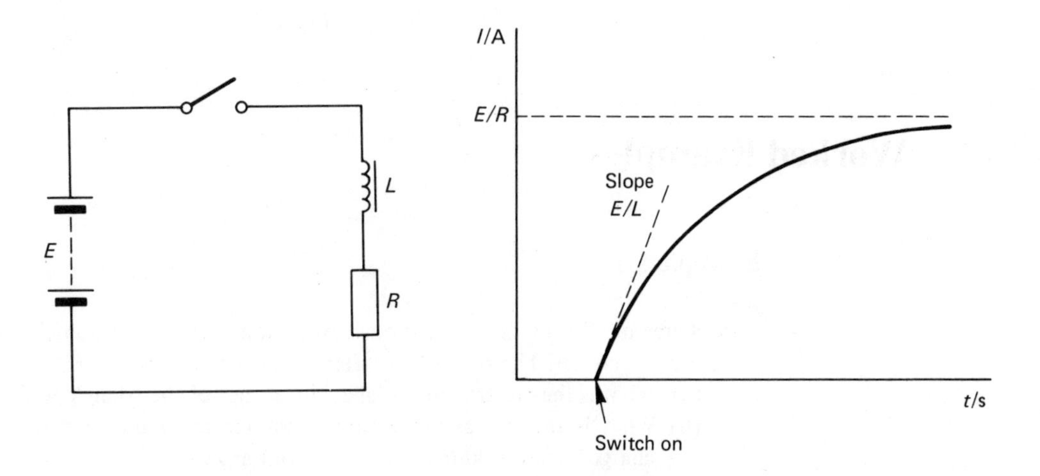

Figure 5.3

graph. The initial slope of the graph (dI/dt) is equal to E/L at switch on and the final current is E/R.

Large transformers can be very efficient, 99%; the small power losses occur as resistive, eddy current and magnetisation heating.

5.5 Alternating Currents

Cathode ray oscilloscopes are used to study a.c. phenomena. The Y-input is calibrated in volts per division (V div^{-1} or mV div^{-1}) and the time-base in seconds per division (s div^{-1} or ms div^{-1} or μs div^{-1}). You should be able to interpret c.r.o. traces.

The relationships between the applied p.d. V, the charge Q and the current I for a capacitor when $V = V_0 \sin 2\pi ft$ are shown in Fig. 5.4. The current is a quarter of a cycle, $\pi/2$ rad, out of phase with the p.d. as $Q = CV$ and $I = dQ/dt$. The current leads the p.d.

In similar graphs for a pure inductor, the current lags behind the p.d. by a quarter of a cycle.

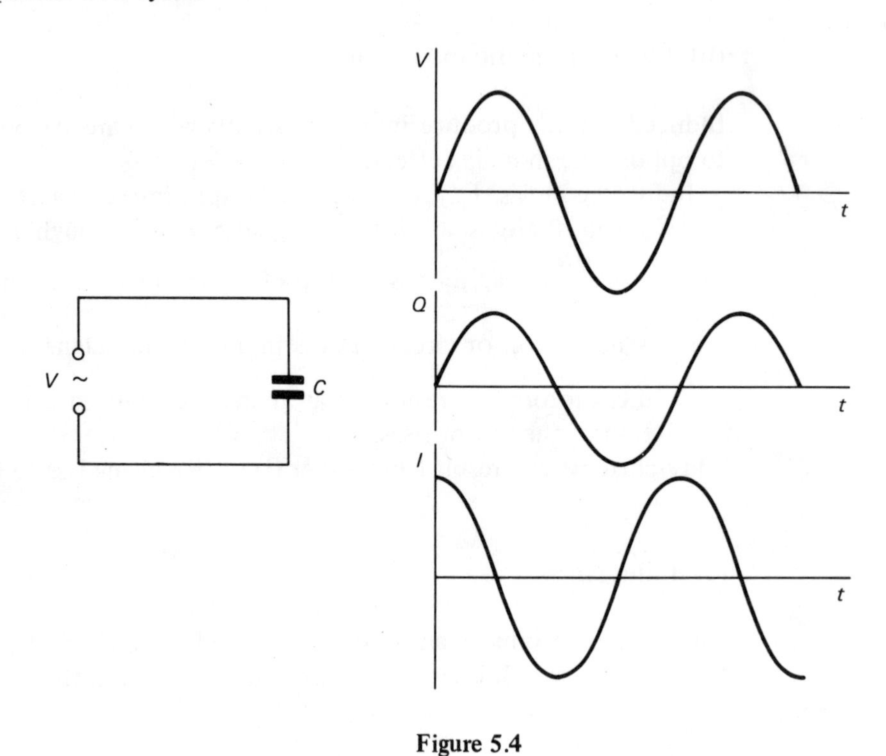

Figure 5.4

5.6 Worked Examples

Example 5.1

A charge of +2.0 μC is placed at one corner of an equilateral triangle and a charge of −2.0 μC at another corner. The sides of the triangle are 1.5 m long.

(a) What is the electric potential at the corner of the triangle without a charge? [2]

(b) What is the size and direction of the electric field strength at the corner without a charge? Draw a sketch to explain your answer. [4]

(c) Calculate the p.d. which would have to be applied between two parallel plates 3.0 mm apart to cause the same electric field as that in (b). [2]

Solution 5.1

(a) Electric potential a distance r from a point charge Q is

$V = Q/4\pi\epsilon_0 r$

The corner of the triangle is the same distance from both charges and they have opposite signs so the total potential will be zero.

Potential is a scalar quantity and so can be added in a simple manner. This is why it is so important and useful.

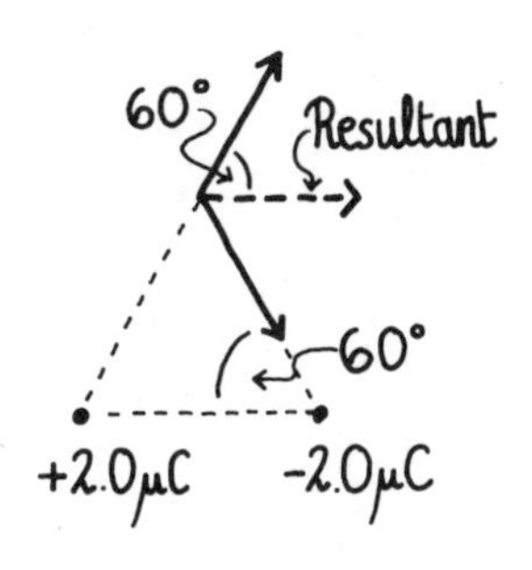

(b) E due to + charge $= \dfrac{1}{4\pi\epsilon_0}\dfrac{Q}{r^2} = \dfrac{2.0 \times 10^{-6}\ \text{C}}{4\pi\,(8.9 \times 10^{-12}\ \text{F m}^{-1})\,(1.5\ \text{m})^2}$

$= 7.9 \times 10^3\ \text{N C}^{-1}$, or V m^{-1}

Because of the 60° angle the vector addition is simple. The resultant E field is the same size as its components and it is parallel to the line joining the charges.

(c) For a uniform field between parallel plates which are at a p.d. V,

$$E = V/d$$

Therefore $V = (8.0 \times 10^3\ \text{V m}^{-1})\,(3.0 \times 10^{-3}\ \text{m}) = 24\ \text{V}$

Example 5.2

(a) Measurements show that there is at the Earth's surface an electric field of strength about 100 V m^{-1} directed vertically downwards. To what surface density of charge does this correspond? Is the sign of this surface charge positive or negative? [3]

(b) The radius of the Earth is 6.4×10^6 m. What will be the electric field strength 3.0×10^6 m above the surface of the Earth? [2]

Solution 5.2

(a) $E = \sigma/\epsilon_0$ since the electric field near the surface is effectively uniform the field lines being almost parallel.

We use the same approximation for the gravitational field close to the Earth. We treat it as having a constant value of about 10 N kg^{-1}.

The surface density of charge, $\sigma = \epsilon_0 E$,

$\sigma = (8.9 \times 10^{-12}\ \text{F m}^{-1})\,(100\ \text{V m}^{-1}) = 8.9 \times 10^{-10}\ \text{C m}^{-2}$

The direction of an electric field is the direction in which positive charge would move. The field is towards the surface so the surface charge must be negative to attract positive.

(b) Electric field strength obeys an inverse square law, taking distances from the centre of the Earth.

Distance of point above surface to centre, $r = (6.4 + 3.0) \times 10^6$ m
$= 9.4 \times 10^6$ m

Electric field strength there $= \left(\dfrac{r_E^2}{r^2}\right)(100\ \text{V m}^{-1})$

$= \left(\dfrac{6.4}{9.4}\right)^2 (100\ \text{V m}^{-1}) = 46\ \text{V m}^{-1}$

This assumes the charge is evenly spread over the whole Earth.

Example 5.3

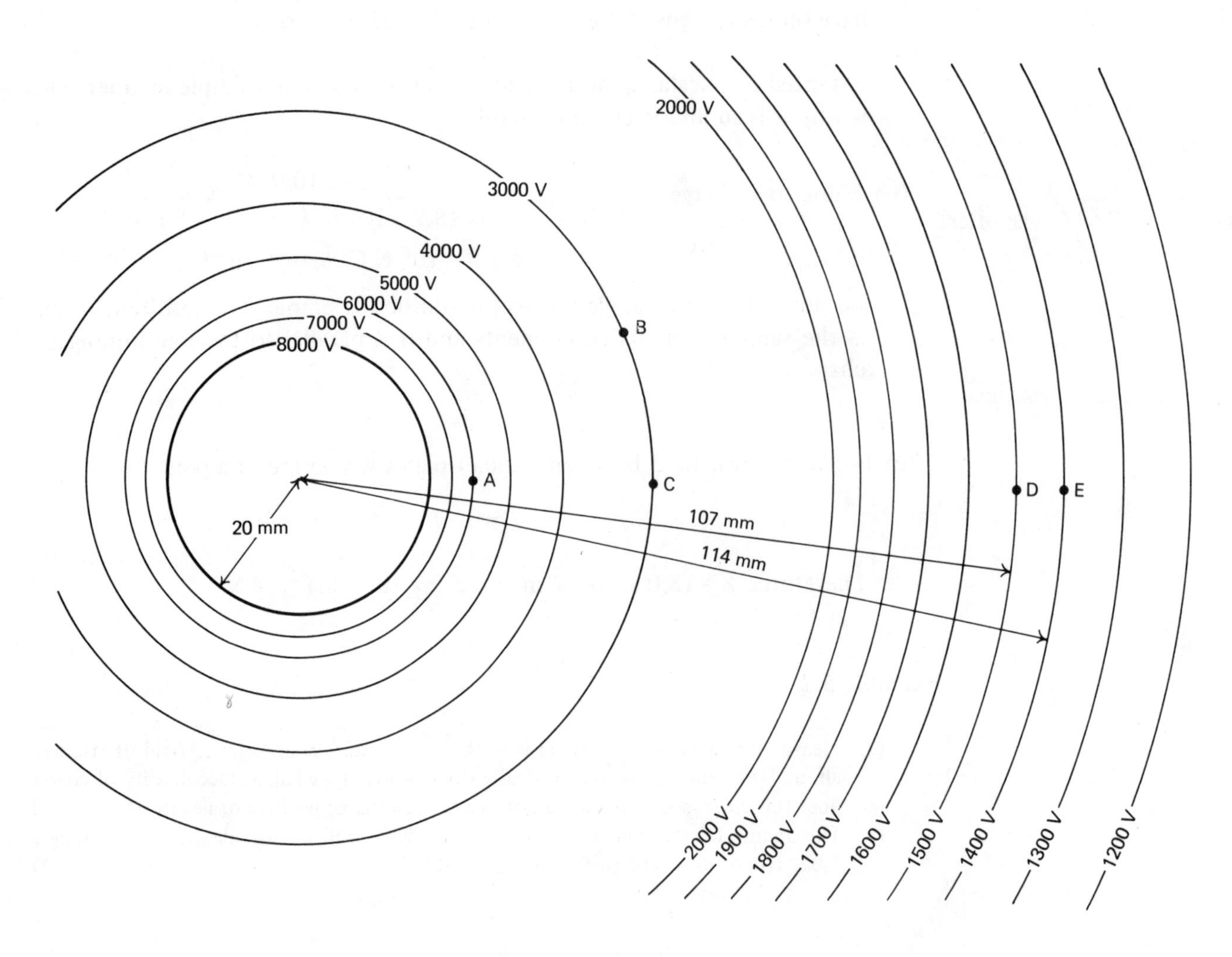

The figure shows equipotential lines around a conducting sphere of radius 20 mm which is charged to a potential of 8000 V. The lines are shown every 1000 V above 2000 V, and every 100 V below 2000 V. The sphere is a long way from other objects.

(a) Is the charge on the sphere positive or negative? [1]

(b) How much energy is required to move a 8.0 nC charge from B to A? [2]

(c) How much energy is required to move a 6.0 nC charge from B to C? [2]

(d) (i) What is the potential difference between D and E?

(ii) How much energy would be required to take 1.0 nC from E to D?

(iii) Estimate the average force acting on a 1.0 nC charge between E and D.

(iv) Estimate the electric field strength between E and D.

(v) Use your answer to (iv) to estimate the charge on the sphere. [10]

(e) An identical sphere with the same charge is brought close to the first sphere so that their centres are 80 mm apart.

(i) What is the potential of the point midway between them?

(ii) How far from the spheres will the potential be 1.0 V? [5]

Solution 5.3

(a) The charge is positive because work is done in bringing unit charge (positive) up to the sphere.

Remember that despite the many analogies between electrical and gravitational fields there is a fundamental difference: mass attracts mass, but like charges repel.

(b) Energy = (6000 V – 3000 V) (8.0 x 10^{-9} C) = 2.4 x 10^{-5} J
(c) The energy is zero because B and C are on the same equipotential line.
(d) (i) 1500 V – 1400 V = 100 V
(ii) Energy = (100 V) (1.0 x 10^{-9} C) = 1.0 x 10^{-7} J
(iii) Energy transformed by working = force x displacement

$$\text{Therefore force} \approx \frac{1.0 \times 10^{-7}\ \text{J}}{(0.114\ \text{m} - 0.107\ \text{m})} = 1.4 \times 10^{-5}\ \text{N}$$

(iv) $$\text{Electric field strength} = \frac{1.4 \times 10^{-5}\ \text{N}}{1.0 \times 10^{-9}\ \text{C}} = 1.4 \times 10^{4}\ \text{N C}^{-1}$$

The equipotentials are fairly evenly spaced so it is possible to treat it like a uniform field.

(v) Since $E = Q/4\pi\epsilon_0 r^2$, $Q = 4\pi\epsilon_0 r^2 E$, so
$Q = 4\pi\ (8.8 \times 10^{-12}\ \text{F m}^{-1})\ (0.11\ \text{m})^2 (1.4 \times 10^4\ \text{N C}^{-1})$
$= 1.9 \times 10^{-8}$ C

It is also possible to calculate the charge using the fact that the potential a distance r from a point charge is given by $V = Q/4\pi\epsilon_0 r$ and so $Q = 4\pi\epsilon_0 rV$ $= 4\pi\epsilon_0$ (0.20 m) (8000 V).

(e) (i) The potential 40 mm from a lone sphere is 4000 V and since potentials add as scalars the potential 40 mm from both spheres is 8000 V.

In fact the potential will be a little lower owing to some redistribution of the charge on the surface of the conducting spheres.

(ii) The 1.0 V line is clearly a long way from the two spheres and so it is possible to treat them as a single doubly charged sphere.

Distance is inversely proportional to potential and proportional to the charge.

$$\text{Distance} = \frac{2\ (8000\ \text{V})\ (0.02\ \text{m})}{1.0\ \text{V}} = 320\ \text{m}$$

Example 5.4

Define magnetic flux density (B). [2]

(a) Describe how the variation in B along the axis both inside and outside a long solenoid carrying a current may be investigated.

Sketch a graph showing the results you would expect to obtain. [9]

(b) The diagram represents a long solenoid in which a steady current is flowing. An

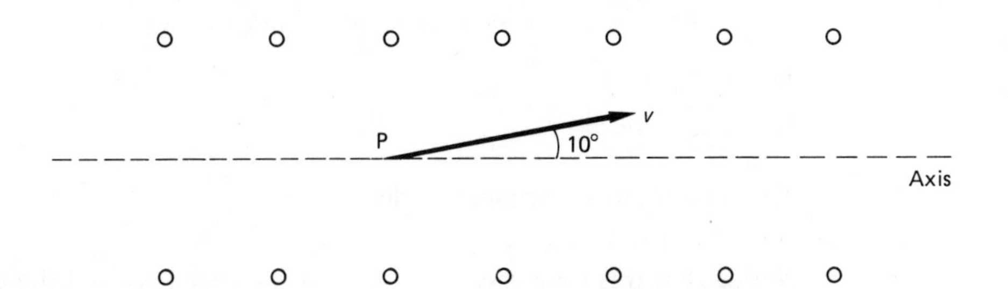

electron is emitted at P with an initial velocity v in the direction shown. By considering the components of v (i) perpendicular to the axis, and (ii) parallel to the axis, deduce the path the electron will follow. If the value of v is 2.0×10^6 m s^{-1} and the flux density at P is 3.0×10^{-4} T, at what distance along the axis will the electron next cross it? [7]

The specific charge of an electron, e/m_e, may be assumed to be 1.8×10^{11} C kg^{-1}.

(L)

Solution 5.4

B indicates the strength of a magnetic field and is a vector quantity because it has direction as well as size. In the S.I. its unit the tesla is defined by the equation

$$B = \frac{F}{Il}$$

where F is the force on a conductor of length l carrying a current I at right angles to the magnetic field B.

(a) Supply the solenoid with alternating current and connect a small search coil with many turns to a c.r.o. Move the coil along the axis with its plane at right angles to the axis of the solenoid. The height of the c.r.o. trace will be proportional to the strength of the magnetic field. The changing magnetic field induces an e.m.f. in the coil and since the frequency is fixed the e.m.f. depends entirely on B.

It is also possible to use d.c. in the solenoid and detect the steady field with a Hall probe.

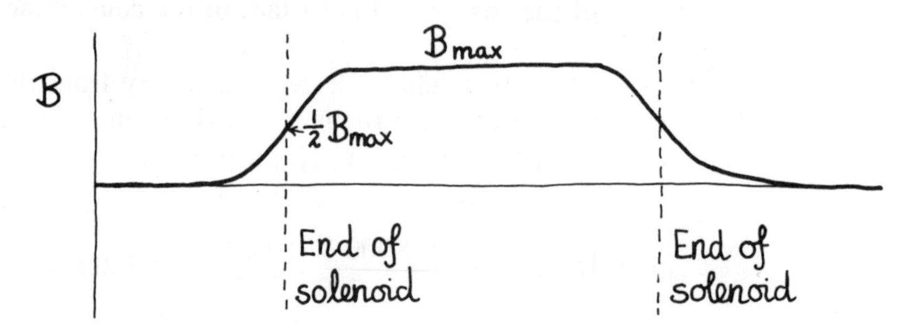

(b) The velocity perpendicular to the axis is $v \sin \theta$. The electron will experience a force at right angles to its motion and at right angles to the field.

The numbers are inconvenient so it is better to use letters until the end. Sometimes you will have to use letters in an unfamiliar way in formulas such as $v \sin \theta$ instead of v.

It will start to move in a circle around the field lines. Its velocity along the axis will be constant and equal to $v \cos \theta$, because this motion is unaffected by the field.

The speed around the circle is $v \sin \theta$ and so the centripetal force

$= Be\,(v \sin \theta) = mr\omega^2$ by Newton's second law.

But $r\omega = v \sin \theta$ $\therefore$ $Ber\omega = mr\omega^2$

Hence $\omega = Be/m$

The time to do a complete circle $= \dfrac{2\pi}{\omega} = \dfrac{2\pi m}{Be}$

Notice that this time is independent of the speed around the circle.

The distance along the axis in this time

$$= (v \cos\theta)\,\frac{2\pi m}{Be} = \frac{(v\cos\theta)\,2\pi}{Be/m} = \frac{(2.0 \times 10^6\ \text{m s}^{-1})\,(\cos 10°)\,2\pi}{(3.0 \times 10^{-4}\ \text{T})\,(1.8 \times 10^{11}\ \text{C kg}^{-1})}$$

$$= 0.23\ \text{m}$$

If θ is small $\cos\theta \approx 1$ and so all the electrons will end up in the same place at the same time a distance 0.23 m along the axis of the solenoid even though the small angle θ may vary. This is the basis of an experiment to measure e/m.

Example 5.5

(a) The flow of electric current in a metal wire is due to the movement of conduction electrons.
  (i) What is meant by the term conduction electrons?
  (ii) Under what circumstances will the movement of the electrons produce current flow? [4]

(b) The diagram shows a length L of conductor of cross-sectional area A. The conductor contains charged particles free to move from left to right as shown.

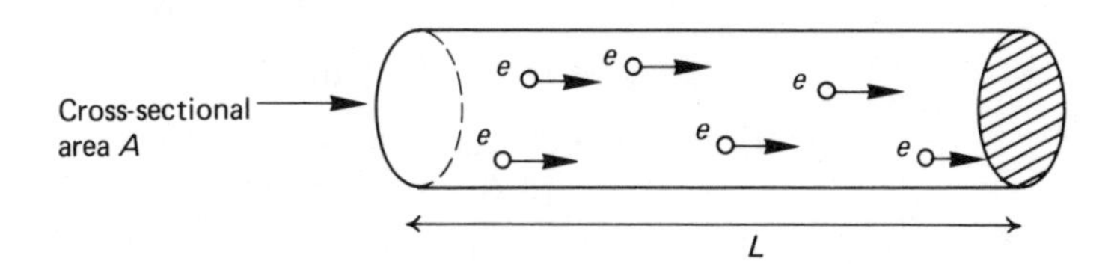

  (i) If there are n such particles per unit volume each with charge e, derive an expression for the total charge of these particles in the length of conductor.
  (ii) If the particles are each moving with a drift velocity v in the direction shown, write down an expression for the time taken for all the particles to pass through the shaded area.
  (iii) Use your answers to (i) and (ii) to show that the current I flowing in the conductor is given by the equation:

$$I = nAve \qquad [5]$$

(c) A current of 1.0 A flows in a 1.0 m length of copper wire of cross-sectional area 1.0 mm^2.
  (i) Given that each cubic metre of copper contains 9.0×10^{28} conduction electrons each of charge -1.6×10^{-19} C, calculate the average drift velocity of the electrons.
  (ii) If the wire is placed at right angles to a magnetic field of 1.0×10^{-2} T, calculate the average force experienced by each electron due to the field.
  (iii) What is the total force experienced by the wire due to the field? [8]

(d) Materials which are good conductors of charge are usually also good conductors of heat. Explain in terms of electron motion how heat may be conducted through a metal bar. [4]

(AEB 1984)

Solution 5.5

(a) (i) Most of the electrons are bound to a particular nucleus. But each atom in a metallic conductor contributes one or two that are free to wander through the metal. These are conduction electrons.

(ii) When a p.d. across a metal causes an electric field the conduction electrons will drift in one direction and produce a current.

The drift speed, calculated in (c) below, is very small compared with the speed, $\approx 10^6$ m s^{-1}, at which the electrons move between collisions with the lattice ions.

(b) (i) AL is the volume of the cylinder and so the number of charges is nAL. The total charge is therefore $nALe$.

(ii) Even those on the far left must pass through the shaded area so the time for all the particles is

$$\frac{\text{maximum distance}}{\text{speed}} = \frac{L}{v}$$

(iii) $$\frac{\text{Total charge}}{\text{time for all charge to flow}} = \text{current}$$

$$I = \frac{nALe}{L/v} = nAve$$

(c) (i) $$v = \frac{I}{nAe} = \frac{1.0 \text{ A}}{(9.0 \times 10^{28} \text{ m}^{-3})(1.0 \times 10^{-6} \text{ m}^2)(1.6 \times 10^{-19} \text{ C})}$$

$$= 6.9 \times 10^{-5} \text{ m s}^{-1}$$

(ii) Force on a charged particle = BQv

$= (1.0 \times 10^{-2} \text{ T})(1.6 \times 10^{-19} \text{ C})(6.9 \times 10^{-5} \text{ m s}^{-1})$

$= 1.1 \times 10^{-25}$ N

(iii) Number of charges in length = nAL

$= (9.0 \times 10^{28} \text{ m}^{-3})(1.0 \times 10^{-6} \text{ m}^2)(1.0 \text{ m})$

$= 9.0 \times 10^{22}$

Force on the wire = $(9.0 \times 10^{22})(1.1 \times 10^{-25} \text{ N})$

$= 1.0 \times 10^{-2}$ N

The final answer is to 2 s.f. if one keeps the numbers in the calculator all the way through. It can be checked using $F = BIL$ $= (1.0 \times 10^{-2} \text{ T})(1.0 \text{ A})(1.0 \text{ m}) = 1.0 \times 10^{-2}$ N

(d) Electrons at the hotter end have more k.e. and so higher speeds. They move randomly and diffuse along the bar faster than electrons from the colder end. They lose energy in collisions with atoms and gradually warm up the whole bar.

Example 5.6

A scientist wishes to study the effect of magnetic fields on a delicate biological specimen by measuring its temperature with a sensitive thermometer.

He places the specimen in a small glass tube at the centre of a solenoid.

(a) What information would he need to calculate the field at the centre of the solenoid? [2]

(b) What might he do to ensure that any temperature change in the specimen was not caused by the heating effect of the current in the coil? [2]

(c) The length and diameter of the solenoid had already been decided as had the fact that it should be wound in a single layer of turns each touching the next. He had 0.2 mm and 0.4 mm diameter wire available for the windings.

(i) How would his choice of wire affect the current required to give a particular magnetic field? [2]

(ii) How would his choice of wire affect the heat caused by that current in the wire? [2]

Solution 5.6

(a) He will need to know the length of the solenoid and the number of turns. From this he can calculate the number of turns per metre. He also needs the current and the value of the permeability of free space, μ_0, which is defined by the definition of the ampere.

(b) He could place the sample in a vacuum flask which would not affect the field, but would prevent it being heated. Or he could use as a control another specimen in a non-inductively wound coil of the same size.

(c) (i) The thinner wire would give twice as many turns per metre so the current would be half that required with the thicker wire.

(ii) The resistance of the thinner wire will be eight times that of the thicker wire because its cross-sectional area is reduced by a factor of 4, and its length is doubled.

The power in an ohmic resistor is given by I^2R. If I in the thinner wire is half that in the thicker wire, I^2 will be four times less in the thinner wire. Since R is eight times bigger, the power in the thinner wire will be twice that in the thicker wire.

Example 5.7

(a) Lenz's law of electromagnetic induction states that the direction of an induced current is such that its effect opposes the change in magnetic flux giving rise to the current. Describe how you would demonstrate the above law using the following apparatus:
an air cored solenoid, a centre zero galvanometer, connecting wires, a magnetic compass, a bar magnet and a dry cell. [7]

Lenz's law is consistent with the law of energy conservation. By referring to the experiment you have described explain why this is so. [3]

(b) In an experimental arrangement shown in the diagram a short bar magnet is dropped and falls through a coil. As the magnet enters the coil the meter is seen to deflect to

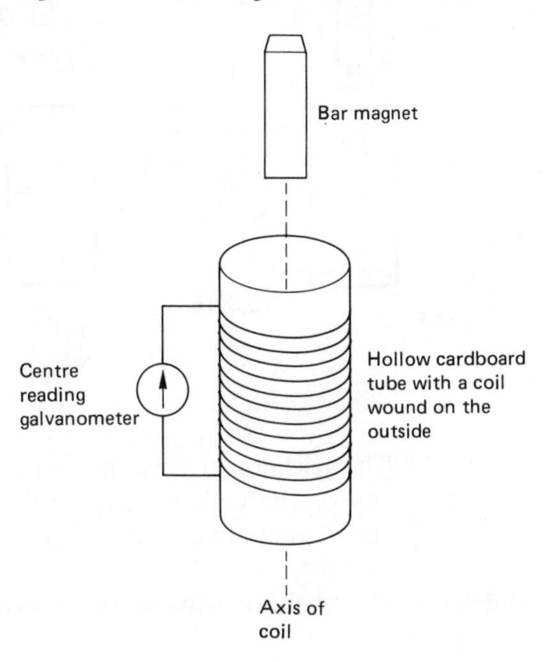

the right. What deflections would you expect as the magnet passes
(i) through the centre, and then
(ii) out of the coil
assuming the axis of the magnet remains vertical?

If the magnet were to be caught at the centre of the coil and then projected upwards, what deflection would the meter show as the magnet passes
(iii) out of the top of the coil?

Explain the reason for your answers in each case. [8]
(L)

Solution 5.7

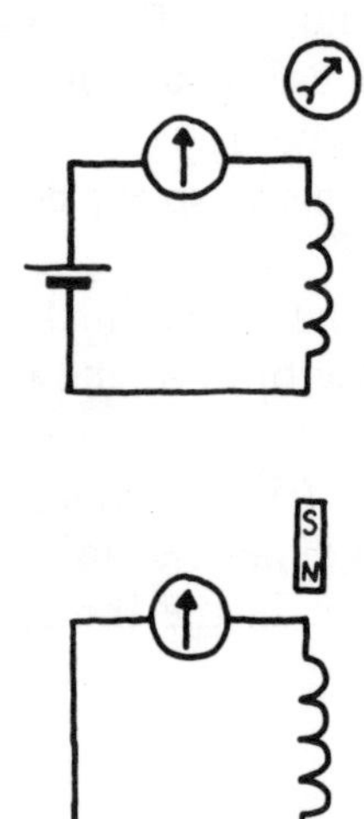

(a) Connect the solenoid, cell and galvanometer as shown. Use the compass to find out which end of the solenoid is N when the current is in a particular direction.

Then remove the cell from the circuit, and check with the compass which end of the magnet is N, in other words north seeking.

Move the magnet towards the coil and see which way the current flows. From the previous experiment it is known which will be the N end.

It will be found that as an N pole approaches the coil that end of the coil becomes N owing to the induced current and so opposes the approach of the magnet and the further increase in flux.

As the coil is moved away, the induced current is in the opposite direction. This causes an attraction between coil and magnet and so opposes the change in flux by preventing it falling as quickly as it might.

If the flux caused by the induced current helped the magnet on its way, the extra motion would produce more help and so there would be mechanical energy coming from nowhere.

(b) (i) As the magnet passes through the central region of the coil there will be no induced current since the total magnetic flux through the coil is not changing.

(ii) As the magnet comes out, the flux will be changing and by Lenz's law the galvanometer will move to the left since the field of the coil is now attracting the magnet.

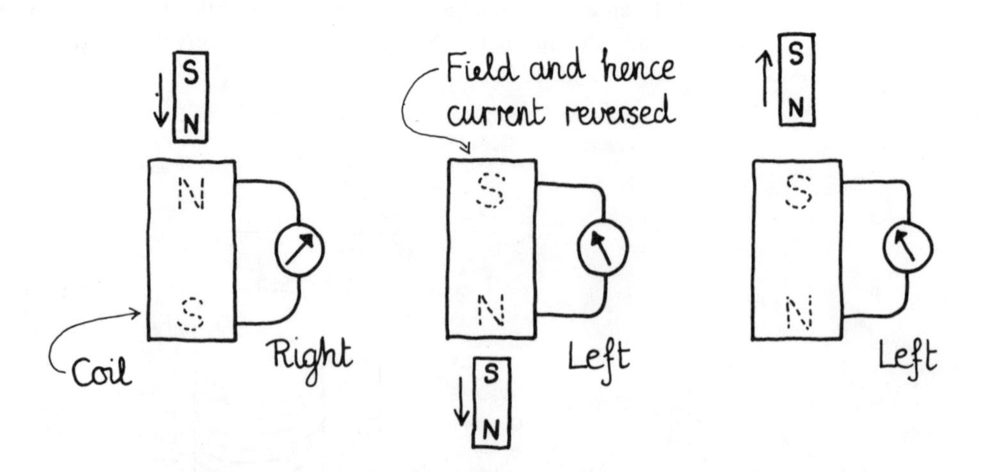

(iii) The galvanometer will move to the left as the magnet comes out of the top of the coil, since again the flux from the coil and the flux from the magnet are in the same direction.

Quick diagrams can save many words of explanation.

Example 5.8

(a) Describe, with the aid of a simple diagram, how a current carrying coil can be connected to a d.c. power supply so that when placed in a uniform magnetic field it will behave as a simple motor. [2]

(b) A d.c. motor uses an electromagnet to produce the field. The coils of the magnet are in series with the armature of the coil as shown.

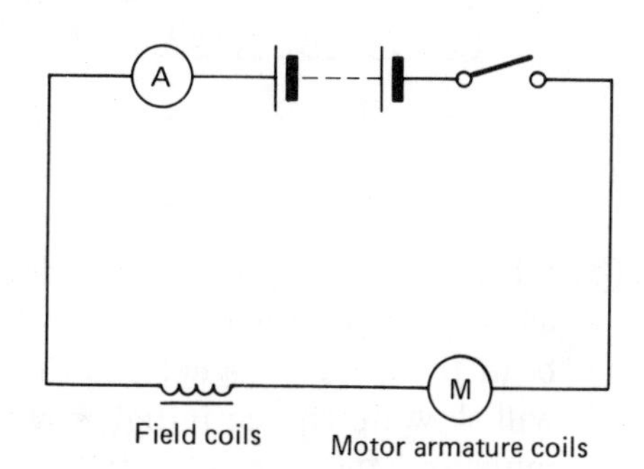

Describe and explain what happens to the current when the switch is closed and a flywheel driven by the motor gradually reaches full speed. [3]

(c) The round cross-section copper bar rolls down the two parallel copper rails as shown in the diagram. Assume that there is good electrical contact but very little friction between the bar and the rails and that the rails are connected by a wire at the top of the slope.

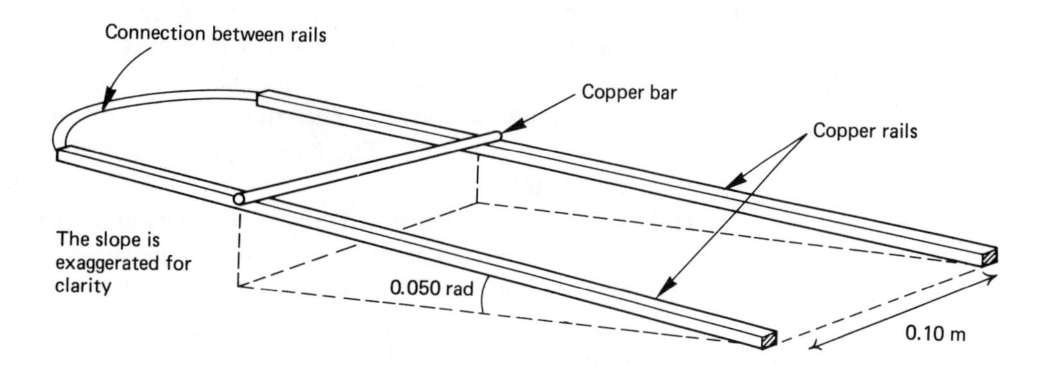

Describe the motion of the bar

(i) when there is no magnetic field,

(ii) when there is a strong vertical magnetic field,

(iii) when there is a strong magnetic field parallel to the bar. [6]

(d) The rails in (c) are at an angle of 0.050 rad (about 3°) to the horizontal and there is a vertical magnetic field strength of 2.0×10^{-2} T. The length of the bar is 0.10 m and is of cross-sectional area 2.0×10^{-6} m^2. Its density is 9.0×10^3 kg m^{-3}. The resistivity of copper is 1.6×10^{-8} Ω m. You may assume that the rails and connecting wires are very thick compared with the bar and that their resistance may be ignored.

(i) Calculate the component of the weight of the bar parallel to the rails in N. [2]

(ii) Calculate the resistance of the bar. [2]

(iii) Calculate the maximum speed of the bar as it rolls down the rails. [5]

You may assume that sin (0.050 rad) = 0.050.

Solution 5.8

(a) Unless the current in the coil is reversed every half turn the coil will not continue to rotate. The commutator shown overleaf makes rotation possible.

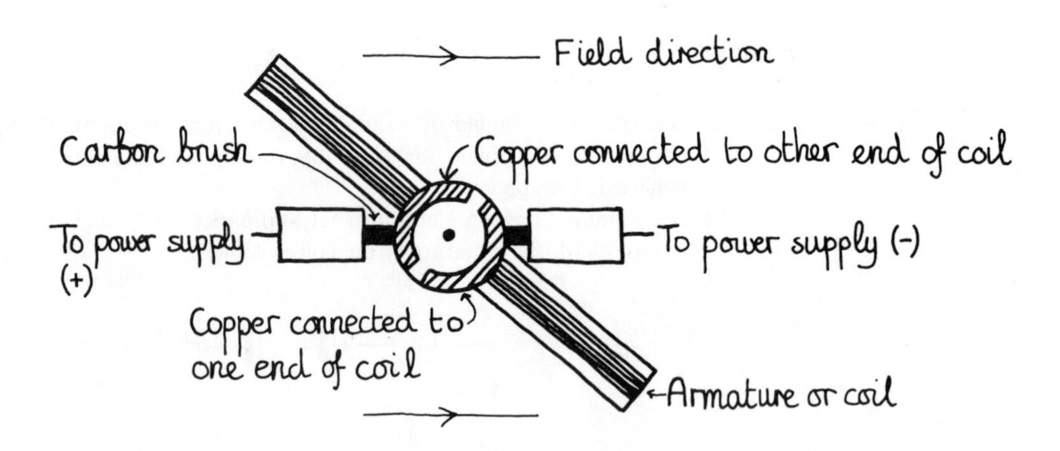

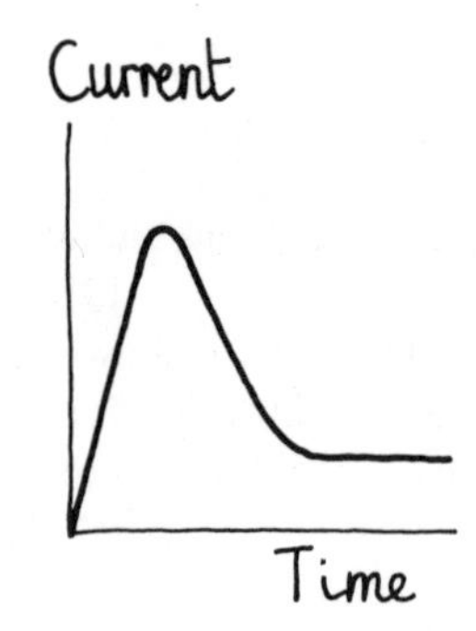

(b) When the motor is not moving no e.m.f. is being induced in the motor as the coil is not cutting the field caused by the electromagnet. However both the electromagnet and the armature will have self-inductances and this will slow down the initial growth of current. The current will reach a maximum and then begin to fall as the motor speeds up and acts as a dynamo opposing the flow of current. The current will reach its lowest value when the flywheel is up to its highest speed. Then the torque of the motor will be equal and opposite to the torque caused by frictional forces. Were there no friction the current would fall to zero as the e.m.f. induced in the motor balanced that of the battery. At any stage the current would go up and down over a cycle of the motor's rotation, but this might not be indicated on a sluggish ammeter.

(c) (i) The bar would accelerate uniformly as it rolled down the slope.

(ii) The bar would gradually get up to a maximum speed where the magnetic force on the current induced in the bar as a result of cutting the field lines would balance the component of the force of gravity parallel to the rails. The bar would be in equilibrium and stay at this maximum speed.

(iii) The bar would not cut any field lines and so there would be no induced current. The motion would be the same as in (i).

(d) (i) Mass of bar = $(0.10 \text{ m})(2.0 \times 10^{-6} \text{ m}^2)(9.0 \times 10^3 \text{ kg m}^{-3})$
$= 1.8 \times 10^{-3}$ kg

Component of weight parallel to rails = $0.050 \times$ weight
$= (0.050)(10 \text{ N kg}^{-1})(1.8 \times 10^{-3} \text{ kg})$
$= 9.0 \times 10^{-4}$ N

If you get confused between cos and sin draw a sketch or think what would happen if the angle were 0. The force would be zero if the rails were level and sin θ is zero when the angle is 0.

(ii) Resistance $= \dfrac{\rho l}{A} = \dfrac{(0.10 \text{ m})(1.6 \times 10^{-8} \ \Omega \text{ m})}{(2.0 \times 10^{-6} \text{ m}^2)} = 8.0 \times 10^{-4} \ \Omega$

(iii) The e.m.f. is equal to
$Blv = v\,(0.10 \text{ m})(2.0 \times 10^{-2} \text{ T})$ where v is the maximum speed.
Current through bar = e.m.f./resistance

$$\therefore \quad F = BIl = \frac{(2.0 \times 10^{-2} \text{ T})\, v\, (0.10 \text{ m})(2.0 \times 10^{-2} \text{ T})(0.10 \text{ m})}{(8.0 \times 10^{-4} \ \Omega)}$$

$= 9.0 \times 10^{-4}$ N (approximately parallel to the rails since θ is small)

Therefore $v = \dfrac{(9.0 \times 10^{-4} \text{ N})(8.0 \times 10^{-4} \ \Omega)}{(2.0 \times 10^{-2} \text{ T})^2 (0.10 \text{ m})^2} = 0.18 \text{ m s}^{-1}$

Example 5.9

A coil of many turns is wound on a soft iron core and is connected in series with a 2 V cell, an ammeter and a switch. A neon lamp, which requires a potential difference of about 200 V across it before it lights, is connected across the terminals of the switch.

When the switch is closed, the ammeter reading rises only slowly and the neon lamp does not light. When the switch is opened, the neon lamp lights momentarily. Explain qualitatively each of these observations, and suggest how the observations might change if the coil were air-cored. [7]

(AEB 1983)

Solution 5.9

When a circuit is described in a question using only words draw a quick circuit diagram for your own benefit.

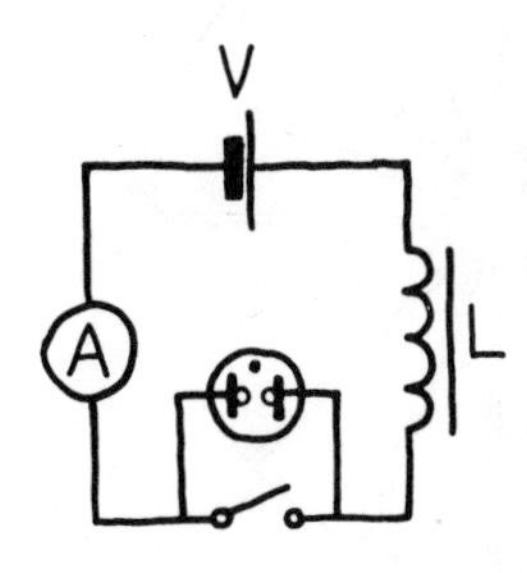

When the switch is closed the current rises slowly because the changing current in the coil causes an induced e.m.f. which is in the opposite direction to the e.m.f. of the cell. The coil's e.m.f. opposes the growth in current. The e.m.f. induced in the coil falls to zero when the current reaches its maximum steady value. The final current depends on the e.m.f. of the cell and the total resistance of the circuit.

When the switch is opened the current rapidly falls and the collapsing magnetic field in the coil induces a high e.m.f. This voltage appears across the neon bulb and being greater than 200 V causes it to light. But the effect is short-lived and the light turns off again quickly.

An air cored coil will have a lower inductance and so the current will rise more quickly. The lamp will probably not light noticeably on opening the switch because the energy stored in the air cored coil is far less.

Petrol engines use this principle to produce the sparks.

Example 5.10

A circuit consists of an air-cored coil, a lamp and an a.c. source all in series. A solid iron core is introduced into the coil.

(a) Explain
  (i) why the lamp is dimmer after the core has been introduced,
  (ii) why the core becomes hot.

(b) What would be the effect on the heating of the core if
  (i) it were laminated with the plane of laminations parallel to the axis of the core,
  (ii) it were laminated with the plane of laminations perpendicular to the axis of the core? [7]

(AEB 1983)

Solution 5.10

(a) (i) The iron core increases the inductance of the coil by making the changing magnetic field in it larger. The reactance of the inductor is increased and so the current is reduced. Thus the bulb in series with the coil is dimmer.

(ii) The changing magnetic field in the coil induces eddy currents in the conducting iron core. These currents cause the core to become warm.

(b) (i) The eddy currents are induced perpendicular to the changing magnetic field which is parallel to the axis of the coil. Thus laminations parallel to the axis tend to prevent their flow and reduce the heating effect.

(ii) Laminations perpendicular to the axis of the core will be parallel to the eddy currents and thus have no effect on them. The core will heat up as much as a solid one.

Example 5.11

The diagram below shows two thin copper discs both of which are able to spin freely about a horizontal axis. Disc B is identical to disc A except for the numerous slits cut through its edge and towards its centre.

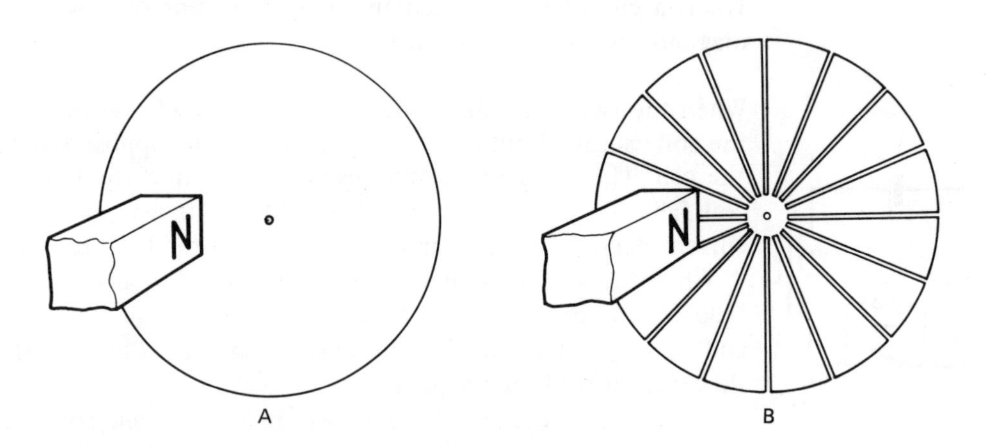

Describe and account for the behaviour of each disc when, set spinning at the same speed, one pole of a strong magnet is held close to the side of the disc and near its edge without touching it. [5]

(L)

Solution 5.11

Disc A will slow down and stop much more quickly than disc B.

Whenever a conductor moves through a magnetic field currents are induced. In the copper disc they are known as eddy currents because they swirl around like water eddies. The slits in disc B prevent large eddy currents developing.

The induced currents in the magnetic field cause a force on the disc and by Lenz's law this force tends to stop the change producing it, that is it slows down the disc. Owing to the slits B is slowed down much less rapidly by this effect; the currents in B are much smaller.

Example 5.12

A cathode ray oscilloscope has its Y-sensitivity set to 10 V cm^{-1}. A sinusoidal input is suitably applied to give a steady trace with the time-base so that the electron beam takes 0.01 s to traverse the screen. If the trace seen has a total peak-to-peak height of 4.0 cm and contains 2 complete cycles, what is

(a) the r.m.s. voltage, and

(b) the frequency

of the input signal? [5]

(L)

Solution 5.12

(a)

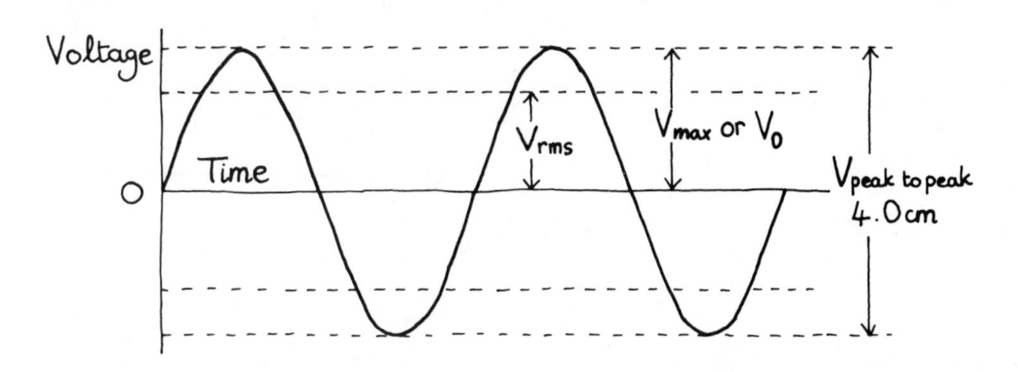

It is almost always worth drawing a sketch and putting the information from the question on it.

Peak-to-peak voltage = (4.0 cm) (10 V cm^{-1}) = 40 V
$\therefore V_{max}$ = 20 V
$V_{rms} = V_{max}/\sqrt{2}$ = 20 V/$\sqrt{2}$ = 14 V

(b) 2 cycles take 0.01 s so the period T = 0.005 s
The frequency, $f = 1/T$ = 1/0.005 s = 200 Hz

There are various ways in which the information might be given to you in a question like this. Read it very carefully.

Example 5.13

A voltage source can produce either a sinusoidal or a square wave alternating voltage as shown in the diagrams.

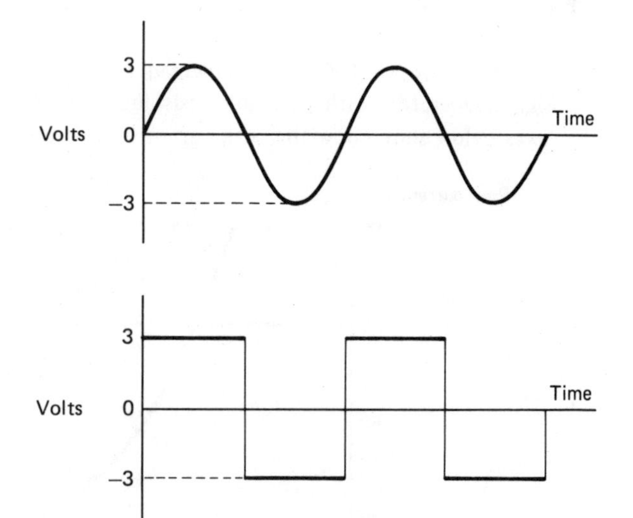

In each case the peak voltage is 3.0 V. Each voltage is applied in turn to a non-inductive resistor of resistance 10.0 Ω.

(a) Sketch graphs to show how the power dissipated in the resistor varies with time over two complete cycles of each wave form.
(b) Calculate the average power dissipated in each case. [7]

(AEB 1983)

Solution 5.13

(a)

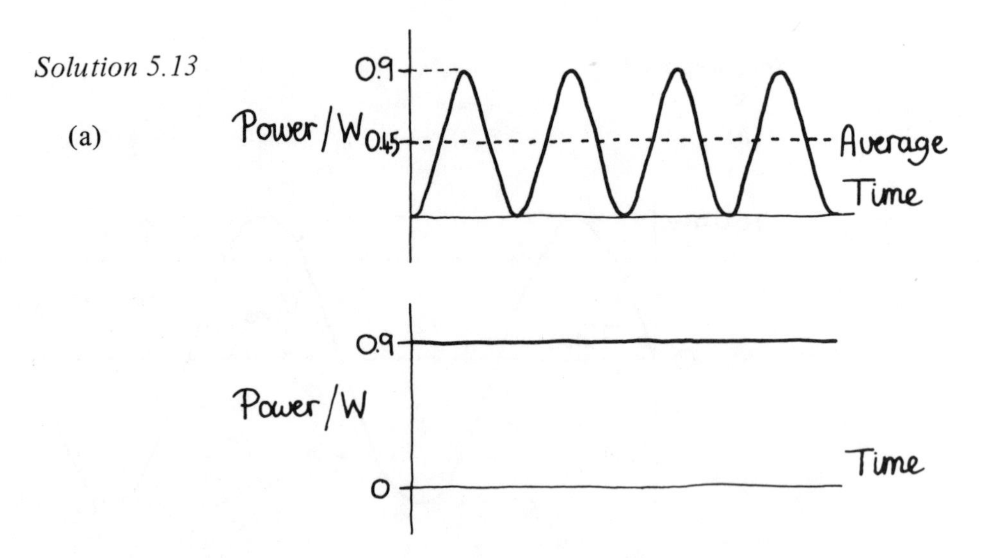

The upper graph is a sinusoidal wave centred on 0.45 W.

If your drawing is not good, be sure to label it so that the examiner knows you meant to draw a sine wave.

The maximum current, $I_{max} = V_{max}/10\ \Omega = 0.30$ A, and because the load is non-inductive it occurs at the same instant as the maximum voltage.
The maximum power = (3.0 V) (0.30 A) = 0.90 W

(b) The average value of the sinusoidal wave is half its maximum value so the average power = 0.90 W/2 = 0.45 W
The average power of the square wave is clearly 0.90 W.

Some square wave generators produce a voltage which alternates between a positive value and zero. Look carefully to see which type of square wave is in the question.

Example 5.14

A resistor, R, of 100 Ω is connected to a variable frequency sinusoidal voltage supply (a signal generator) with negligible resistance in its output circuit. The graph marked I_R on the axes below shows how the peak current in the resistor varies with the frequency of the signal

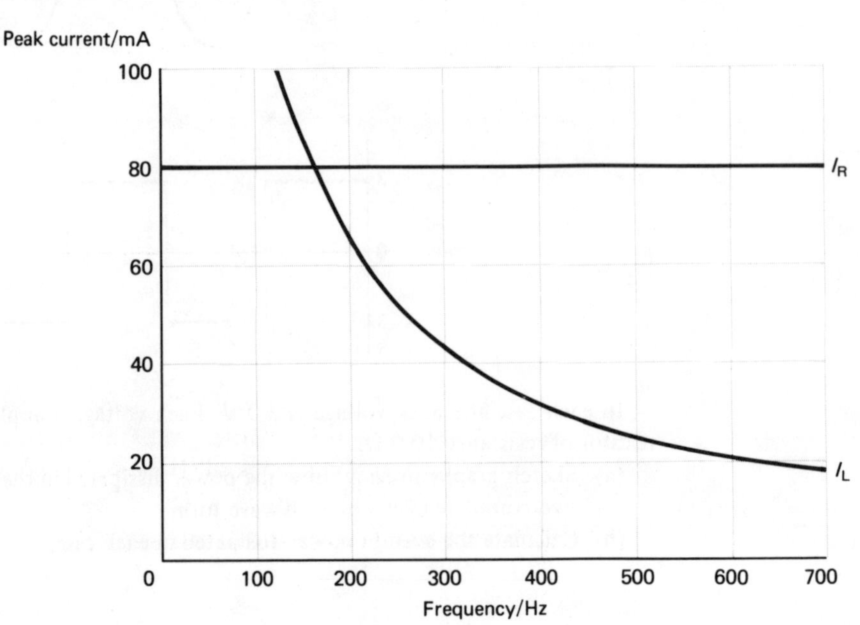

generator. The resistor is replaced by a pure inductor, L, of 0.10 H. The graph marked I_L shows how the peak current in the inductor varies with the frequency of the signal generator.

(a) (i) Calculate the peak voltage of the signal generator. What tells you it is constant over the range of frequencies on the x-axis?

(ii) Calculate the average power dissipated in the resistor. [4]

(b) (i) Given the equation $V = L\, dI/dt$ for a pure inductor, calculate the maximum value of the rate of change of current during a cycle.

(ii) What is the value of the current in the inductor when dI/dt is at its maximum value?

(iii) Use information from the graph to show that I_L is inversely proportional to frequency. [5]

(c) The inductor and the resistor are now connected in series with the same signal generator. Say what you can about the phase of:

(i) the current in L compared with the current in R;

(ii) the p.d. across L compared with the p.d. across R. [4]

(d) The components are still connected in series as described in (c). Without doing detailed calculations sketch a graph showing how the peak current in R varies with frequency over the range from 0 to 700 Hz. Explain the form of your graph. [5]

Solution 5.14

(a) (i) Peak voltage = (peak current) × (resistance)

$= (0.080 \text{ A}) (100 \ \Omega) = 8.0 \text{ V}$

The peak current is constant so the peak voltage must be constant too.

(ii) For a sinusoidal current in a resistor

Average power = peak power/2 = (0.080 A) (8.0 V)/2 = 0.32 W

At any instant, power = p.d. × current, by the very definitions of the quantities.

Average power for sinusoidal currents in resistors = $\dfrac{I_{max} \times V_{max}}{2}$

Using r.m.s. values: average power = $I_{rms} \times V_{rms}$

(b) (i) Maximum rate of change of current,

$$\frac{dI_{max}}{dt} = \frac{V_{max}}{L} = \frac{8.0 \text{ V}}{0.10 \text{ H}} = 80 \text{ A s}^{-1}$$

(ii) When the value of a sine curve is changing most rapidly its actual value is zero.

The current is zero when dI/dt is a maximum.

A sine graph is steepest where it crosses the x-axis.

(iii) If $I_L \propto 1/f$ then $I_L f$ should be constant.

Checking with three sets of data from the graph,

$60 \times 215 = 12\,900$

$32 \times 400 = 12\,900$

$18 \times 700 = 12\,600$

The products are the same given the number of significant figures to which the graph can be read. Hence I_L is inversely proportional to f.

If checking a graph explain the principle of your check and use at least three pairs of data.

(c) (i) The phase difference between the currents is zero.

Whatever the components, if they are in series the current through each must be the same at every instant. Kirchhoff's laws apply to a.c. too!

(ii) Since the p.d. across L is 90° ($\pi/2$ rad) out of phase with the current through it and since the p.d. across a resistor is in phase with the current through it, the p.d. across L is 90° out of phase with the p.d. across R.

(d)

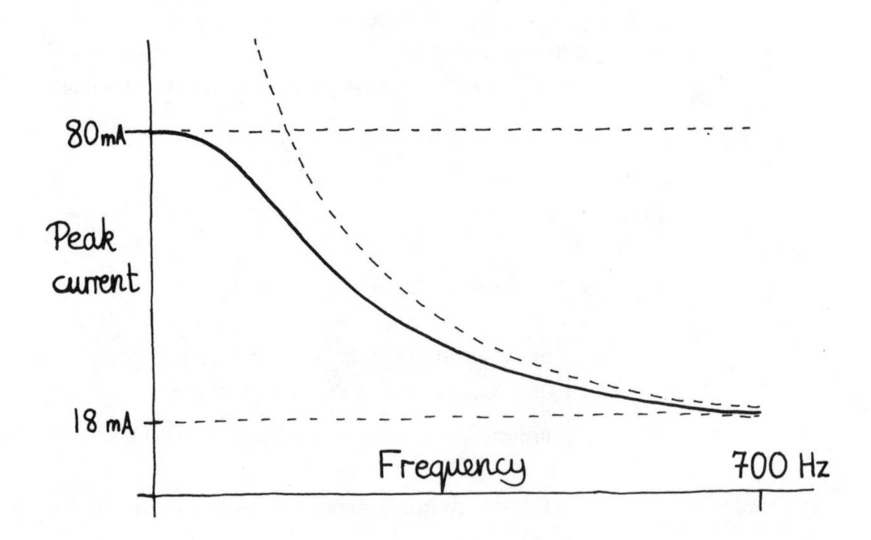

At low frequencies only the resistance of R matters and so the peak current is 80 mA. At high frequencies the reactance of L is more important than R, so the peak current is just a little less than the current through L when by itself.

Sketching in the graphs from the question will show the examiner that you know what you are doing.

5.7 Questions

Question 5.1

Electric charges Q, Q, Q and $-Q$ are set at four corners A, B, C, D, respectively, of a square of side d. D is the furthest corner from A.

(a) Deduce an expression for the electric potential at A produced by the charges at B, C and D.

(b) What is the magnitude and direction of the resultant force on the charge at A produced by the charges at B, C and D? [8]

Question 5.2

+6000 V

The diagram shows two parallel plates, the upper plate being at a potential of 6000 V above that of the lower plate which is at earth potential. Copy the diagram (exact accuracy is not required) and sketch field lines and equipotentials both between the plates and for a short distance beyond either end. State how equipotentials are related to field lines. Indicate the region in which the electric field strength is uniform and state how your diagram shows this to be the case. [7]

(AEB 1983)

Question 5.3

A piece of demonstration equipment used in a school laboratory consists of two long thin strips of aluminium foil side by side. They are parallel and hang vertically. Each strip has electrical connections top and bottom so a current can flow through it, but the strip is not so taut it cannot move sideways slightly.

The tops of the strips are connected together and the bottom of one is attached to the positive terminal of a car battery and the bottom of the other to the negative terminal.

(a) (i) Do the strips move together slightly or move apart? [1]
(ii) Explain why the strips should experience a force. [3]
(iii) Sketch a horizontal cross-section of the magnetic field around the strips. [2]

(b) What will be the effect on the force on the strips of:
(i) moving them twice as far apart?
(ii) doubling the current? [2]

(c) Calculate the approximate force on the central 10 cm of the strip of foil if the strips are 1.0 m long and 2.0 cm apart and the current is 20 A. [3]

(d) How does your calculation in (c) relate to the definition of the ampere? [2]
(Take $\mu_0 = 4\pi \times 10^{-7}$ H m^{-1}.)

Question 5.4

A proton of mass m and charge Q travels with speed v in a circle of radius r in a magnetic field of flux density B.

Write down an expression linking the magnetic force and the proton's acceleration and hence show that the time T for one revolution of the proton is independent of its energy. [4]

Question 5.5

In an experiment to determine the strength of the horizontal component of the Earth's magnetic field an electric current was passed through a long wire taped to the laboratory bench. Vertically above the centre of the wire, which ran north–south (magnetic), a compass was supported on a wooden stand.

With the compass needle a height 18 mm above the wire, it was found that when the current in the wire was adjusted to a value of 1.4 A, it produced a deflection of exactly 45° in the compass, that is the compass then pointed NE. By varying the current I at a series of heights h of the compass and adjusting I in each case so that the compass needle was deflected through 45° the following results were obtained:

$h/10^{-3}$ m	8	18	23	27	35
I/A	0.8	1.4	2.0	2.5	3.1

(a) Sketch the experimental arrangement using symbols for the electrical circuit. [3]

(b) Draw a graph of I (y-axis) against h (x-axis). [4]

(c) (i) Show that, if the horizontal component of the Earth's magnetic field is B_h, the slope of the graph is equal to $2\pi B_h/\mu_0$.
(ii) Calculate a value for B_h given that $\mu_0 = 4\pi \times 10^{-7}$ H m^{-1}. [5]

(d) Estimate the uncertainty in your value of B_h and suggest how the experiment could be improved to give a better value of B_h. [4]

Question 5.6

The following passage is taken from an article entitled 'Domestic science' by C. J. Myers in *Physics Bulletin*, June 1973.

Ignition failures

One of the simplest everyday acts – that of lighting a gas jet – seems to have caused a lot of people a lot of trouble. Just looking at the amazing number of different gadgets which have been produced gives some idea of the problems. There was the old gas wand which still persists today based on a chemical catalyst, the battery operated hot filament device which burnt out in the flame which had just been lit, the 'caliper device' which showered everyone with sparks and left a very distinctive smell, the simple mains operated filament which of course required another connection, and the foolproof pilot light – foolproof that is until the slot meter ran out in the night or until it was simply blown out. One of the latest in this long line comes from Plessey; it is, of course, transistorized. This gadget will produce a spark within 30 μs which means that the gas is ignited almost immediately, avoiding the massive and dangerous build up of gas which other devices needed. The whole unit is powered by a 1.5 V battery and a transistorized oscillator circuit consisting of a silicon planar transistor and a coil. A second coil steps up the 20 V pulses from the oscillator to 300 V which then charges the main 1.5 μF capacitor.

The oscillator produces some 5 kHz and it takes some 1000 pulses before the capacitor is fully charged. When the charge reaches a predetermined level the current is switched (by a gas filled gap set to breakdown at a preset voltage) through the primary of the second transformer giving up to 15 000 V at the spark gap. The typical energy of each spark is 1 mJ and the device will continue sparking every 30 μs. One interesting point is that the energy supplied at the gap is independent of the state of the battery, thus always ensuring a successful light. The energy is controlled by the rate at which the gas filled gap breaks down and also by the energy stored by the capacitor which is equal to $\frac{1}{2}CV^2$. As the battery voltage decreases the oscillator circuit output decreases also and a greater number of pulses is required to charge the capacitor; the frequency of the pulses therefore decreases but the energy is constant. That is until the battery goes completely flat.

Questions

(a) Draw a circuit diagram to illustrate the various stages of the transistorized gas lighter. You should use symbols for individual components but the oscillator can be represented by a box. Label the turns ratios of the transformers and add other component values wherever possible. [8]

(b) There is a mistake in the article. The device does not 'continue sparking every 30 μs', it is the duration of the sparks which is 30 μs, not their frequency.
Use the passage to show that the frequency of sparking is about 5 Hz. [2]

(c) (i) Calculate the maximum energy stored in the main 1.5 μF capacitor. Suggest why your answer is much greater than the typical energy of each spark.
(ii) Calculate the output power of the sparking system. If the current drawn from the battery is 400 mA, what is the efficiency of the system?
(iii) How does the efficiency of the system change as the battery runs down? [8]

5.8 Answers to Questions

5.1 (a) Using $V = Q/4\pi\epsilon_0 r$ three times and adding the potentials as electric potential is a scalar gives $V_A = Q/4\pi\epsilon_0 d$.

(b) Using $F = Q_1 Q_2 / 4\pi\epsilon_0 r^2$ three times gives a force

$F_A = \dfrac{Q^2}{8\pi\epsilon_0 d^2}(2\sqrt{2} - 1)$ in the direction from D to A but the addition is more complicated as force is a vector.

See also worked example 5.1.

5.2 See the introduction to the chapter (page 122). Parallel field lines or parallel equipotentials indicate a uniform field.

5.3 (a) (i) Unless you know, you need to use the right-hand screw rule and Fleming's left-hand rule.
They move apart.
(b) (i) The force is halved.
(ii) The force is quadrupled.
(c) 4×10^{-4} N
(d) Read up the definition if in doubt.

5.4 Use Newton's second law for the centripetal acceleration of the particle subject to a magnetic force $F = BQv$ to show that $T = 2\pi m/Bq$

This is independent of v. See also worked example 5.4.

This result is the basis for the design of a particle accelerator called a cyclotron. It is limited in use because at high energies the mass *m* of the proton increases – a relativistic effect.

5.5 (a) A three-dimensional sketch of the wire and compass is expected. The circuit could be added or you could draw a separate circuit diagram.
(c) (i) When the compass is at 45° the magnetic field, produced by the current in the wire, is equal to B_h.
(ii) 1.8×10^{-5} T.
(d) The graph points do not form a very good straight line and there is no reason why it should go through the origin. Lines giving slopes of ±5% difference could easily be drawn and possible lines might have slopes with ±10% difference. The uncertainty is thus at least $\pm 0.1 \times 10^{-5}$ T.
Ways of improving the experiment would include methods of getting several readings of I for each value of h.

5.6 **You should not have spent too much time reading and re-reading the passage before tackling the questions.**

(a)

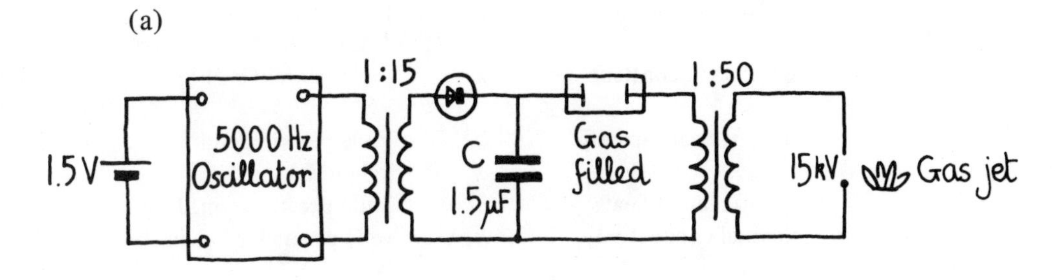

The diode is not mentioned in the passage but is essential if the capacitor is not to discharge between pulses.

(c) (i) 70 mJ (ii) 5 mW, 0.8%

6 Mechanical and Thermal Properties of Matter

6.1 You Should Recognise

Quantity	*Symbol*	*Unit*	*Comments*
area	A	square metre, m^2	$1\ mm^2 = 10^{-6}\ m^2$
pressure	p	pascal, Pa	$1\ Pa = 1\ N\ m^{-2}$
tension	T	newton, N	
tensile stress	σ	pascal, Pa	breaking stress (strength) of copper $\approx$ 200 MPa
linear strain	ϵ	(no unit)	often expressed as %
Young modulus	E	pascal, Pa	$1\ kN\ mm^{-2} = 10^9$ Pa or 1 GPa
temperature	θ, t	degree Celsius, °C	
thermodynamic (absolute) temperature	T	kelvin, K	$\theta/°C = T/K - 273.15$
temperature interval	$\Delta T, \Delta\theta$	kelvin, K	
energy transferred by heating	Q	joule, J	often called simply 'heat'
internal energy	U	joule, J	
mean squared speed of molecules	$\langle c^2 \rangle, \overline{c^2}$	metre squared per second squared, $m^2\ s^{-2}$	
r.m.s. speed of molecules	c_{rms}	metre per second, $m\ s^{-1}$	about 500 $m\ s^{-1}$ for air molecules at s.t.p.
number of molecules	N	(no unit)	
amount of substance	n	mole, mol	often thought of as 'number of moles'
Avogadro number	N_A	per mole, mol^{-1}	a constant = $6.03 \times 10^{23}\ mol^{-1}$
molar gas constant	R	joule per mole kelvin, $J\ mol^{-1}\ K^{-1}$	a constant = $8.31\ J\ mol^{-1}\ K^{-1}$
Boltzmann constant	k	joule per kelvin, $J\ K^{-1}$	a constant = $1.38 \times 10^{-23}\ J\ K^{-1}$
specific heat capacity	c	joule per kilogram kelvin, $J\ kg^{-1}\ K^{-1}$	for water $c = 4200\ J\ kg^{-1}\ K^{-1}$
specific latent heat	l	joule per kilogram, $J\ kg^{-1}$	
thermal conductivity	k, λ	watt per metre kelvin, $W\ m^{-1}\ K^{-1}$	k (metals) $\approx 200k$ (glass or brick)

Not listed are quantities, such as mass and time, which occur in all chapters and which are given in the list on page 7.

6.2 You Should be Able to Use

- For linear extension and compression:

  Young modulus $= \dfrac{\text{tensile stress}}{\text{tensile strain}} = \dfrac{\sigma}{\epsilon}$

  $\sigma = \dfrac{T}{A}$ and $\epsilon = \dfrac{\Delta l}{l_0}$ so that $E = \dfrac{\sigma}{\epsilon} = \dfrac{T}{A} \times \dfrac{l_0}{\Delta l}$

  work done in stretching $W = \text{(average force) (extension)} = F_{av}\Delta l$

  when Hooke's law applies, i.e. when $F = kx$, $W = \frac{1}{2}Fx = \frac{1}{2}kx^2$

- Pressure:

  pressure and force $p = \dfrac{F}{A}$

  pressure change with depth in a fluid $p = \rho g \Delta h$

  work done in compressing a gas at constant pressure $W = p\Delta V$

- Temperature scales:

  Celsius scale with a thermometric property X $\dfrac{\theta}{100°\text{C}} = \dfrac{X_\theta - X_0}{X_{100} - X_0}$

  Kelvin scale from a constant gas thermometer $T = \dfrac{p_T}{p_{tr}} \times 273.16\ \text{K}$

  where p_T and p_{tr} are the pressures at the temperature T and the triple point

- Ideal gases:

  macroscopic (volume V, pressure p, temperature T) $pV = nRT$

  microscopic (mass of molecule m, N molecules) $pV = \frac{1}{3}Nm\langle c^2 \rangle$

  $p = \frac{1}{3}\rho\langle c^2 \rangle$

  temperature and energy $\frac{1}{2}m\langle c^2 \rangle = \frac{3}{2}kT$

- Calorimetry:

  heating/cooling matter without change of state $\Delta Q = mc\Delta\theta$

  changing the state of matter at constant temperature $\Delta Q = ml$

- First law of thermodynamics:

  gain of internal energy $\Delta U = \Delta Q + \Delta W$ where Q is the transfer of energy by heating *to* the gas and W is the work done *on* the gas

- Thermal conduction:

  heat flow rate $\propto$ temperature gradient $\dfrac{dQ}{dt} = -kA\dfrac{d\theta}{dx}$

6.3 Deformation of Solids

Solids transmit forces. Stress–strain graphs for polycrystalline (copper wire), glassy and polymeric (rubber cord) materials are shown in Fig. 6.1. The initial slope of each graph from the origin is the Young modulus of the material. In general the slope of the graph is a measure of the stiffness of the material. Force-extension graphs have the same shape and from them the work done in loading (or unloading) can be found.

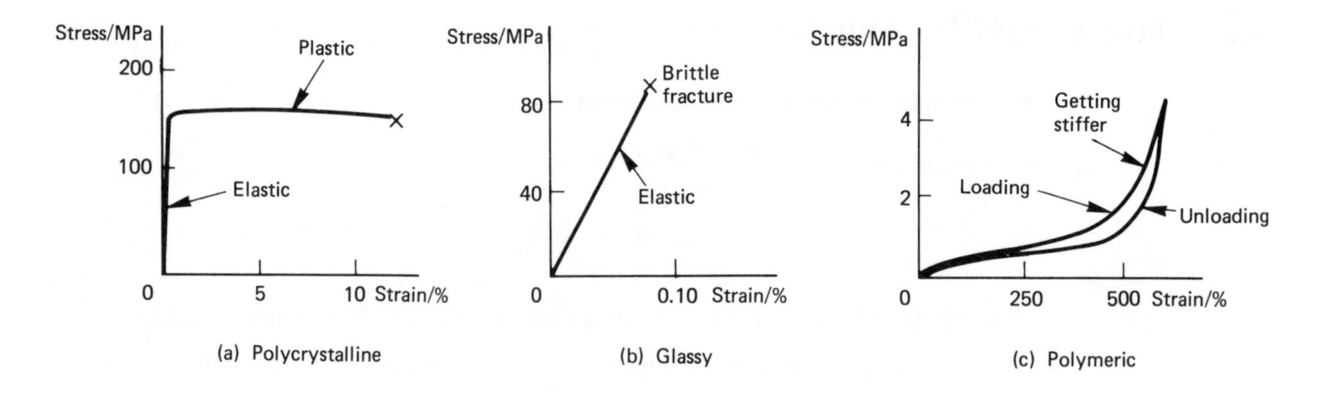

Figure 6.1

6.4 Gases

(a) Macroscopic Properties

Fluids, that is gases and liquids, transmit pressures. A fixed mass of gas at constant temperature obeys Boyle's law, pV = constant. When used at low pressures it defines an ideal gas scale of temperature (the absolute scale), p = (constant) T. These together lead to the ideal gas equation (see above), which applies to many real gases in our normal experiments and calculations.

The behaviour of gases is often described by p–V diagrams (see Fig. 6.2). Sudden expansions and contractions, for which $\Delta Q = 0$, are said to be adiabatic. You should think of graphs like this with the first law of thermodynamics in mind, e.g. when $\Delta W = 0$, then $\Delta U = \Delta Q$, i.e. all the heat goes to raise the internal energy and hence the temperature of the gas.

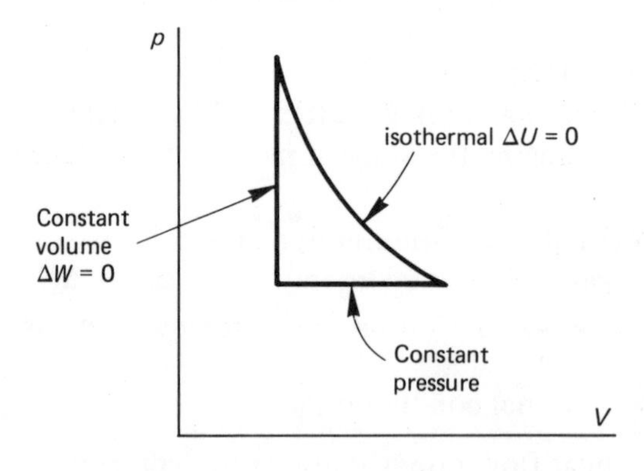

Figure 6.2

(b) Kinetic Theory

A gas exerts a pressure by the bombardment of a surface by its atoms or molecules. By assuming that gas molecules are tiny smooth elastic spheres, the kinetic theory of gases calculates the rate of change of momentum of N molecules each of mass m at the wall of the containing vessel and hence the pressure that the gas exerts. You should be able to reproduce this proof (see worked example 6.5). The molecules move at random with a range of speeds and an average (translational) kinetic energy $\frac{1}{2}m\langle c^2 \rangle$ which is proportional to the absolute temperature of the gas.

6.5 Heat

(a) Temperature Scales

A laboratory thermometer is calibrated to read degrees Celsius (centigrade) using the relationship given on page 145 which is equivalent to the straight line of Fig. 6.3. For different thermometric properties X, e.g. the length of a mercury thread or the resistance of a coil of wire, θ, except at 0°C and 100°C, will be different, though not perhaps by much.

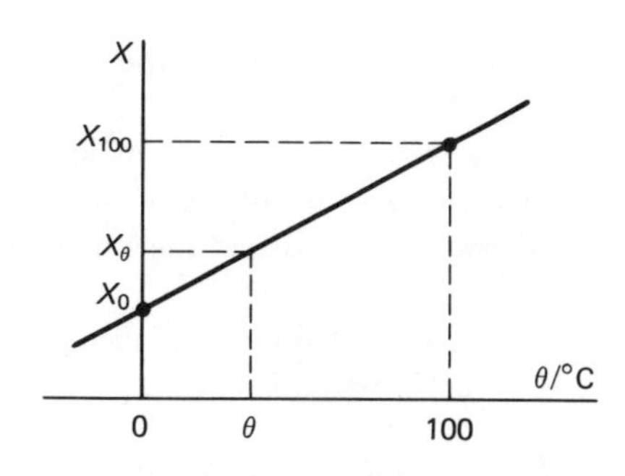

Figure 6.3

(b) Heating Matter

A graph of temperature against time for a substance which starts as a solid and ends as a gas is shown in Fig. 6.4. If the rate of heating is constant then in this case

$$l\,(\text{boiling}) > l\,(\text{melting}) \qquad \text{and} \qquad c\,(\text{solid}) < c\,(\text{liquid})$$

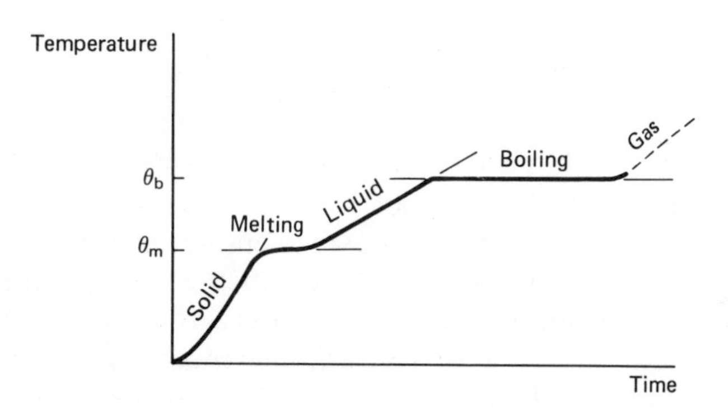

Figure 6.4

(c) Thermal Conduction

The rate of heat transfer through a bar of material depends on its cross-sectional area A, the temperature difference $\Delta\theta$ and the length of the bar l. You can write the equation for thermal conduction to look very like that for electrical conduction when a current $I = \mathrm{d}Q/\mathrm{d}t$ flows in a material of electrical conductivity $\sigma = 1/\rho$ because of a potential difference ΔV:

$$\text{heat} \quad \frac{\mathrm{d}Q}{\mathrm{d}t} = \frac{kA}{l}\Delta\theta \qquad \text{charge} \quad \frac{\mathrm{d}Q}{\mathrm{d}t} = \frac{\sigma A}{l}\Delta V$$

The U-value of a wall is the value of k/l and must be below $1.0 \text{ W m}^{-2} \text{ K}^{-1}$ to satisfy building regulations. When you measure k for a poor conductor you choose a specimen with a small length l and a large cross-sectional area A. This is so that the heat flow is approximately across the specimen and of a measurable size. You must also wait for steady conditions to be reached.

6.6 Worked Examples

Example 6.1

(a) Explain the meaning of the terms ***stress*** and ***strain*** when applied to the deformation of a stretched wire. [2]

(b) Draw a labelled diagram of an apparatus you would use to investigate the way in which the strain of a steel wire depends on the stress applied. How would you use this apparatus to find the Young modulus of the material of which the wire is made? [8]

(c) (i) A load of 60 N is applied to a steel wire of length 2.5 m and cross-sectional area 0.22 mm^2. The Young modulus for steel is 210 GPa. What extension is produced?

(ii) A temperature rise of 1 K causes a fractional increase of 0.001% in the length of a steel wire. If the temperature were to increase by 4 K during the experiment, calculate the change in length of the wire.

(iii) Discuss whether an increase in temperature is important in the experiment you have described to measure the Young modulus of steel. [8]

Solution 6.1

(a) $$\text{Stress} = \frac{\text{tensile force}}{\text{cross-sectional area}}$$

$$\text{Strain} = \frac{\text{increase in length}}{\text{original length}}$$

Explain the meaning really means define the quantities.

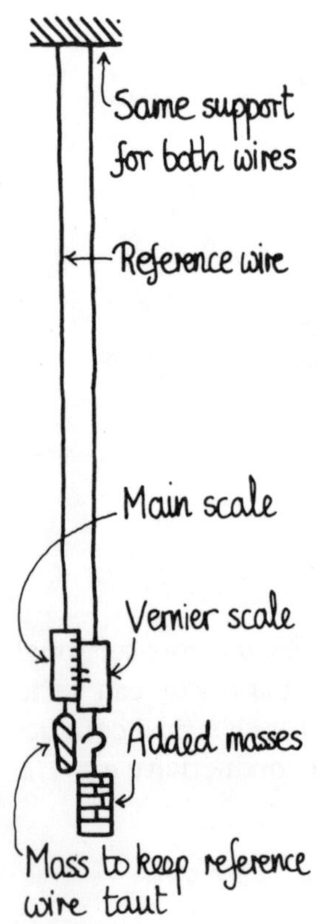

(b) Add enough masses to ensure the wires are taut.

Measure the length of the wire l_0 with a metre rule. Take the measurement from the top to the zero of the vernier.

Add masses, each time recording the vernier reading.

Plot a graph of vernier reading against mass as you do the experiment.

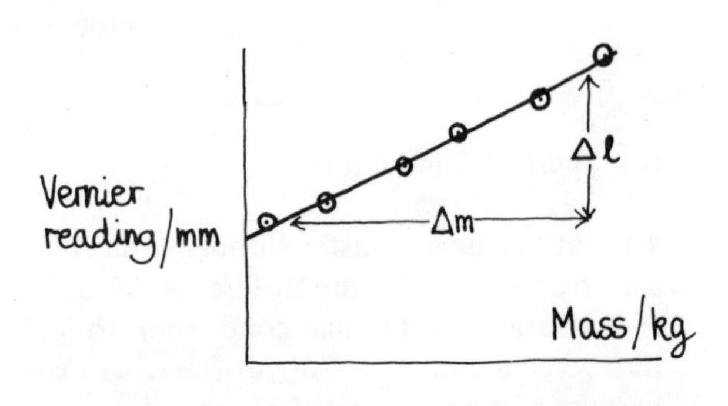

You are of course going to calculate the weights of the masses, but it is best to deal with 'raw' data early on in the experiment.

Measure the diameter of the wire in a few places with a micrometer screw gauge and take an average value with which to calculate the cross-sectional area using $A = \pi d^2/4$.

Unload the wire to check no plastic deformation has taken place.

Calculate the slope of the graph $\Delta l/\Delta m$.

Young modulus, $E = \dfrac{g\Delta m}{A} \times \dfrac{l_0}{\Delta l}$ where $g\Delta m$ is the change in force

$$= g \times \frac{l_0}{A} \times \frac{1}{\text{slope of the graph}}$$

(c) (i) Rearranging the equation

$$\Delta l = \frac{Tl_0}{AE} = \frac{(60\ \text{N})\ (2.5\ \text{m})}{(0.22 \times 10^{-6}\ \text{m}^2)\ (210 \times 10^9\ \text{Pa})} = 3.2 \times 10^{-3}\ \text{m}$$

(ii) 0.001% of 2.5 m = 0.000 01 x 2.5 m = 2.5×10^{-5} m or 25 μm. So for a 4 K rise in temperature the wire increases in length by 4 x 25 μm. This is 100 μm or 0.10×10^{-3} m.

Strains are sometimes given as a percentage, sometimes not. Read the question carefully.

(iii) If the temperature changed during the experiment, both the wire under test and the reference wire would alter in length. Provided they were made of the same material the increase would be the same for both and would not affect the vernier readings. The slight effect on the length of the wire could be ignored in the calculation.

Example 6.2

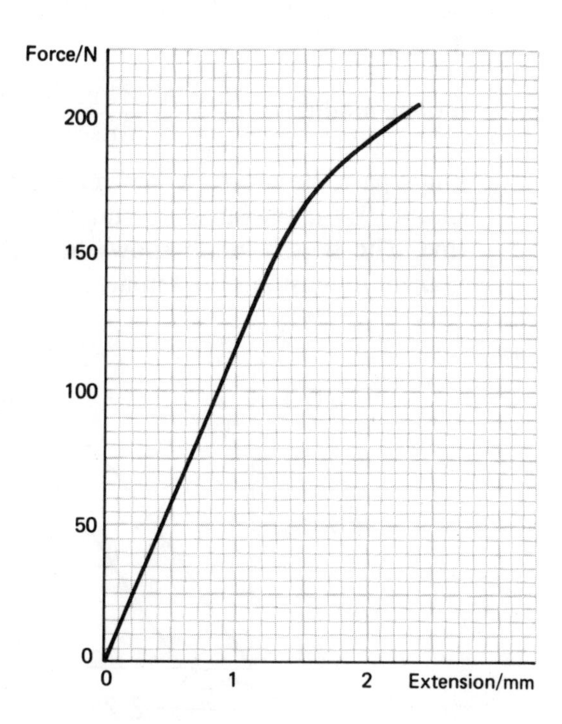

(a) The graph shows how the tensile force in a steel wire varies with extension. Over what range of force does the wire obey Hooke's law? [2]

(b) Using the graph estimate the mechanical work done when stretching the steel wire from no extension to an extension of 1.0 mm. [2]

(c) What information must be obtained from the graph to help calculate the Young modulus for the wire? State what other information about the steel wire is needed and explain how to calculate the Young modulus. [3]

(d) Estimate the permanent extension produced when a load of 200 N is hung on the steel wire and then taken off. Explain your reasoning carefully. [3]

Solution 6.2

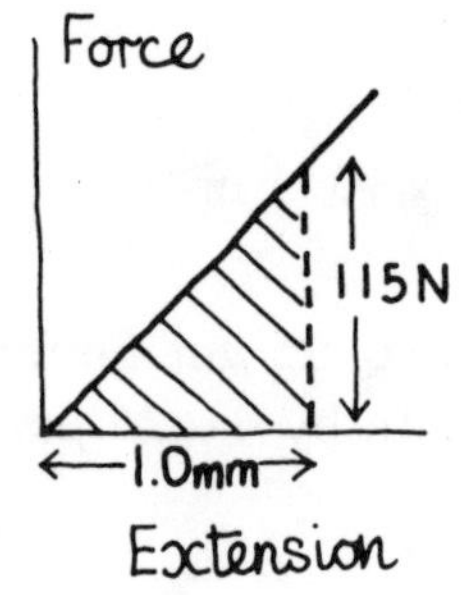

(a) Hooke's law is obeyed when the tension is proportional to the extension. The range is from a force of 0 up to 150 N where the line starts to bend.

(b) Work done (energy) is the area under a force-extension graph. The area is triangular

so work done $= \frac{1}{2}$ (115 N) (1.0 x 10^{-3} m)
$= 0.058$ J

(c) The gradient of the graph tells us the force per unit extension for a particular wire. To find the Young modulus of the metal we need to know its original length and cross-sectional area.

$$\text{Young modulus} = \frac{\text{tensile stress}}{\text{tensile strain}} = \frac{(\text{force}) \times (\text{original length})}{(\text{area}) \times (\text{extension})}$$

Since stress = $\frac{\textbf{force}}{\textbf{area}}$**, and strain =** $\frac{\textbf{extension}}{\textbf{original length}}$

From graph,

$$\frac{\text{force}}{\text{extension}} = \frac{115\ \text{N}}{1.0 \times 10^{-3}\ \text{m}} = 115\,000\ \text{N m}^{-1}$$

Therefore

$$\text{Young modulus} = \frac{115\,000\ \text{N m}^{-1} \times (\text{original length})}{(\text{cross-sectional area})}$$

(d) Once past the Hooke's law region of the graph most of the extension is plastic and if the stress is reduced the wire contracts down another straight line. The sketch graph shows how the graph was used to estimate a permanent extension of 0.4 mm.

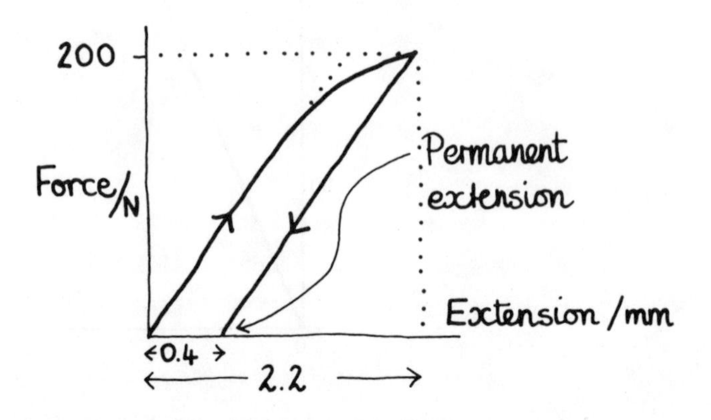

Do not be afraid to draw a sketch graph – it's often easier than explaining using only words.

Example 6.3

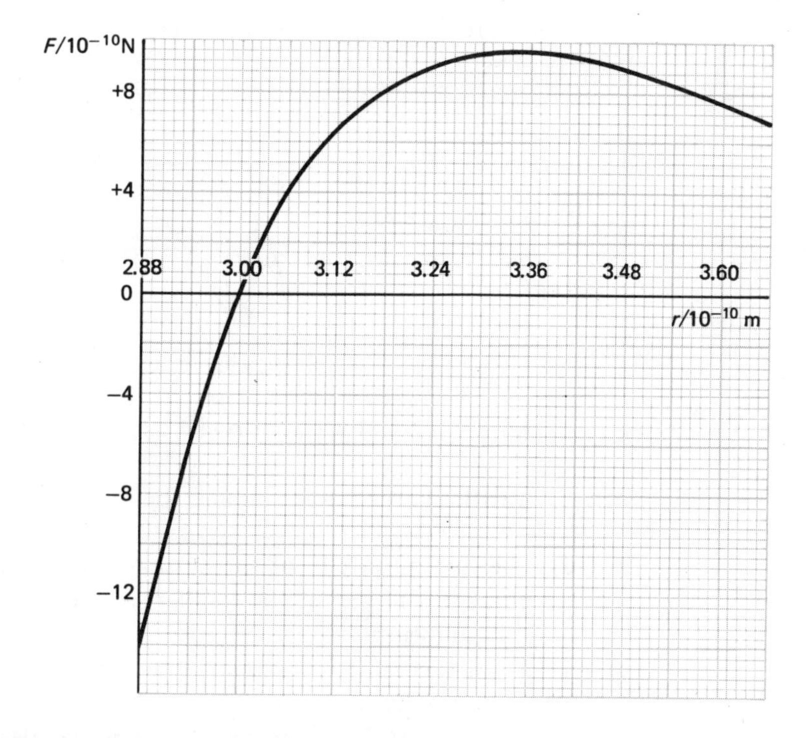

The graph shows how the resultant force, F, between a pair of molecules might vary as the separation, r, between them increases. Positive values of the force indicate that it is an attraction, negative values indicate that it is a repulsion.

(a) Use the graph to determine

(i) the equilibrium separation and the separation at which the attraction force is a maximum,

(ii) the energy required to decrease the separation of the molecules from 3.00×10^{-10} m to 2.88×10^{-10} m, and

(iii) the potential energy of the molecules at a separation of 2.88×10^{-10} m measured with respect to the minimum value of the potential energy of the system. [5]

(b) The model as represented by the graph can be used to help explain the elastic behaviour of solids. Use it to explain

(i) why solids resist being stretched or compressed,

(ii) why Hooke's law is obeyed for small changes in the length of a solid rod,

(iii) why, above a certain value of the tensile stress, a solid will break. [7]

(c) It can be shown that the Young modulus for a solid is approximately given by $E = k/x$ for small displacements, where x is the equilibrium spacing between the molecules and k is the force per unit displacement between the molecules. To what quantity does k relate on the graph? Use the graph to make an estimate of the value of k explaining how you obtained your result. Calculate a value for the Young modulus, for the solid represented by the graph. [6]

(L)

Solution 6.3

(a) (i) In equilibrium $F = 0$ so separation is 3.00×10^{-10} m

Notice 3 significant figures.

Maximum attractive force is at 3.33×10^{-10} m.

Be careful about the strange scale; 10 squares are 0.12×10^{-10} m.

(ii) Since energy = (average force) × (distance), the area 'under' the graph is required.

Treating the area as a triangle.

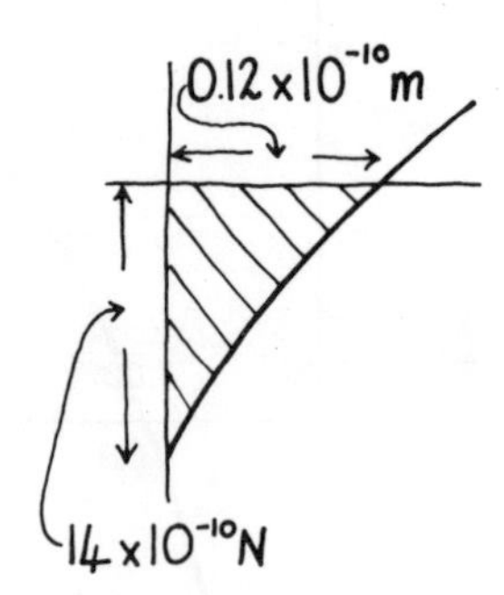

$$\text{Area} = -\tfrac{1}{2}(14 \times 10^{-10}\ \text{N})\ (0.12 \times 10^{-10}\ \text{m}) = -0.8 \times 10^{-20}\ \text{J}$$

A little sketch helps you and the examiner.

(iii) The minimum p.e. of the system is when the molecules are in equilibrium, so the p.e. at the smaller separation is 0.4×10^{-20} J higher than at equilibrium.

(b) (i) A solid consists of many molecules joined together by bonds which are normally in the equilibrium state.

If the solid is stretched the molecules are pulled further apart against an attractive force which resists the extension.

If the solid is compressed the molecules are pushed closer together against a repulsive force which resists the compression.

(ii) Near the equilibrium point of zero force the graph is almost straight and therefore the force is proportional to the difference in length of the bonds from their equilibrium length. So for the whole solid the force is proportional to the extension or compression. This is Hooke's law.

(iii) Eventually the force applied will be sufficient to break the bonds completely. The largest force required to separate the molecules is 10×10^{-10} N so the breaking force will be this times the number of bonds in a cross-section.

In reality solids are not as strong as this model predicts, because of dislocations and impurity atoms.

(c) k is the gradient of the graph.

Using the straightest part of the graph a change in length from 3.00×10^{-10} m to 3.03×10^{-10} m leads to a change in force of 2.00×10^{-10} N

$$\therefore k = \frac{2.0 \times 10^{-10}\ \text{N}}{0.03 \times 10^{-10}\ \text{m}} = 67\ \text{N m}^{-1}$$

$$\text{and } E = \frac{k}{x} = \frac{67\ \text{N m}^{-1}}{3.0 \times 10^{-10}\ \text{m}} = 2.2 \times 10^{11}\ \text{N m}^{-2}$$

Example 6.4

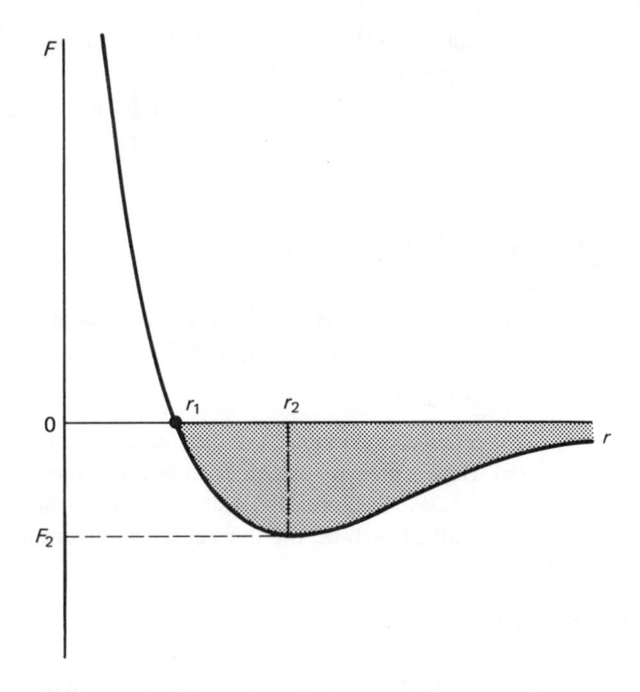

(a) The graph shows the way in which the resultant force F between a pair of atoms in a metal wire depends on the distance r between them. Explain the physical significance of the following features of the graph:

(i) positive values of F,

(ii) r_1, the value of r when $F = 0$,

(iii) the slope of the graph at $r = r_1$,

(iv) r_2, the value of r when $F = F_2$,

(v) the shaded area on the graph. [6]

(b) Predictions of the values of the Young modulus from such graphs are in reasonable agreement with values obtained from experiments with wires, but the breaking strains of wires are normally very much smaller (by a factor of 100 or so) than values predicted from the graphs. Account for the agreement in the former case, and disagreement in the latter, when both are based on the same simple model of forces between atoms. [4]

Solution 6.4

(a) (i) The atoms repel each other for positive F.

(ii) This is the centre-to-centre distance between the atoms in an unstressed metal wire – the equilibrium separation. It is equal to d, the effective diameter of one atom. It is also the separation at which the atoms have minimum potential energy.

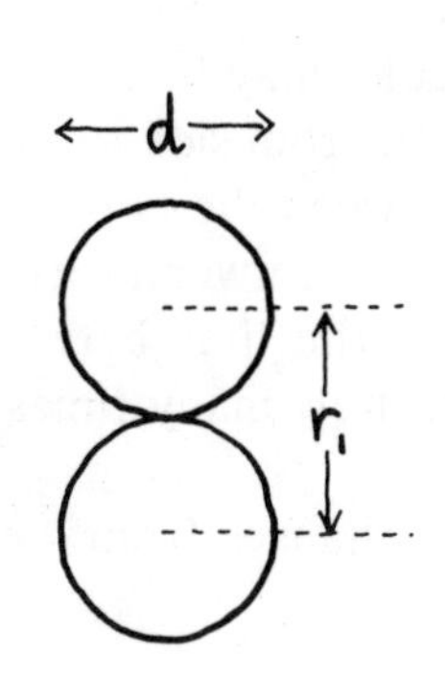

Notice r_1 is not a radius. It is very easy to confuse radii and diameters.

(iii) At $r = r_1$ and for small changes in distance Δr, because the line is nearly straight, F is proportional to Δr, or $F = k\Delta r$. This is Hooke's law for both compression and extension.

(iv) If the atoms are pulled a distance greater than $r_2 - r_1$ apart, they will then separate completely with no increase in force.

(v) The shaded area (assuming it continues to the right until $F = 0$) represents the energy needed to separate the atoms completely. This is related to the latent heat of vaporisation of the metal.

(b) The strain of a metal wire is the same as the fractional extension $\Delta r/r$ of the two atoms. The stress is related to the Hooke's law constant, k, in (a) (iii) and so the Young modulus can be related to k.

Wires break not because the intermolecular force rises to F_2 but because dislocations allow the atoms to move relative to another at much smaller stresses.

Example 6.5

(a) Explain what is meant by an ideal gas. What properties are assumed for the model of an ideal gas molecule in deriving the expression

$$p = \tfrac{1}{3}\rho\overline{c^2}$$

where the symbols have their usual meanings? [5]

(b) How is pressure explained in terms of the kinetic theory of gases?

Describe carefully, using diagrams where necessary, but without detailed mathematical analysis, the steps in the argument used to derive the equation in (a). [8]

(L)

Solution 6.5

(a) An ideal gas obeys the gas laws completely. pV is a constant at constant temperature and p is proportional to the absolute temperature T at constant volume.

The properties of an ideal gas are:

1. A gas consists of a vast number of molecules in random motion.
2. The molecules are very small – most of the gas is empty space.
3. The molecules exert no forces on each other except when in contact.
4. All collisions involving molecules with each other or molecules with the walls of the container are perfectly elastic – no k.e. is lost.
5. The duration of collisions is very short compared with the time between collisions.
6. The molecules have a range of speeds.

The question asks for the properties of the molecule, but it is best to put all the assumptions of the model because it is not entirely clear, for example, whether the speed is a property of the gas or the molecule.

(b) The bombardment of the walls containing the gas causes a force on them. Think of a cube containing the gas. A third of the molecules may be thought of as going up and down, a third in and out and a third from side to side. Think of a single molecule bouncing back and forth between the two walls shown in the diagram. We can find the distance it moves between hitting the left wall twice and so using its speed can calculate the time between collisions with the left wall. From this time we work out how many times it hits the left wall each second.

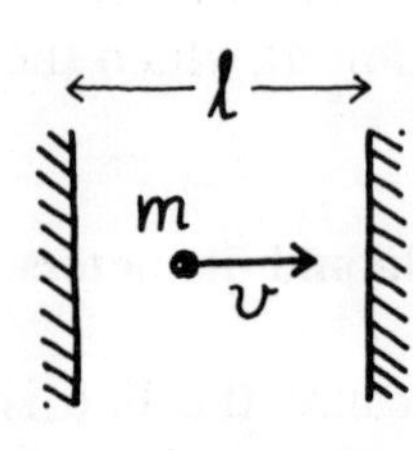

Each time the molecule hits the wall its momentum changes from $-mv$ to $+mv$, a change of $2mv$.

Momentum is a vector quantity.

We can work out the change in momentum per second as the molecule bounces off the left wall. This by Newton's second and third laws will be equal to the force it exerts on the wall.

If we now remember that a third of the molecules are bouncing off the left wall and treat them as having the average speed we can calculate the total force on the wall. We know the area of the wall so we can calculate the pressure in terms of the number of molecules in the box, their mass and average speed and the size of the box. We can write the density in terms of the number, mass and size of the box to get the equation required.

Sometimes a full proof is required. Try producing the full proof using the explanations above and the bare bones of the mathematics below.
Distance between collisions = $2l$

Time between collisions = $\dfrac{2l}{v}$, **frequency of collisions =** $\dfrac{v}{2l}$

Rate of change of momentum = $2mv \times \dfrac{v}{2l} = \dfrac{mv^2}{l}$

Force due to all molecules = $\dfrac{Nm\overline{c^2}}{3l}$

Area of wall is l^2 so pressure = $\dfrac{Nm\overline{c^2}}{3V}$

Density of gas = $\dfrac{\text{mass}}{V} = \dfrac{mN}{V} = \rho$

Finally $p = \frac{1}{3}\rho\overline{c^2}$
The one third of the molecules assumption is a simplification but is widely accepted. Treating all the molecules as going at the average speed is too.

Example 6.6

(a) The kinetic model of an ideal gas is used to help explain the behaviour of real gases. Write down two of the assumptions made about the molecules which make up an ideal gas and outline the experimental evidence supporting one of the assumptions. [4]

(b) A cyclist notices that the end of his aluminium bicycle pump becomes quite hot while he is pumping up his tyres. Explain carefully why the pressure and temperature of air in the pump increase during a single push on the piston. You should mention the motion of molecules in your answer. [4]

(c) A hand pump is being used to inflate a bicycle tyre in which the air is already 400 kPa *above* atmospheric pressure. The initial volume of the air trapped by the piston in the pump is 1.0×10^{-4} m^3 and this air is at atmospheric pressure and 27°C. The volume of air trapped is 2.5×10^{-5} m^3 at the moment when the one-way valve between the pump and the tyre begins to open. Calculate the temperature of the air in the pump at this moment. (Atmospheric pressure = 100 kPa.) [4]

(d) 

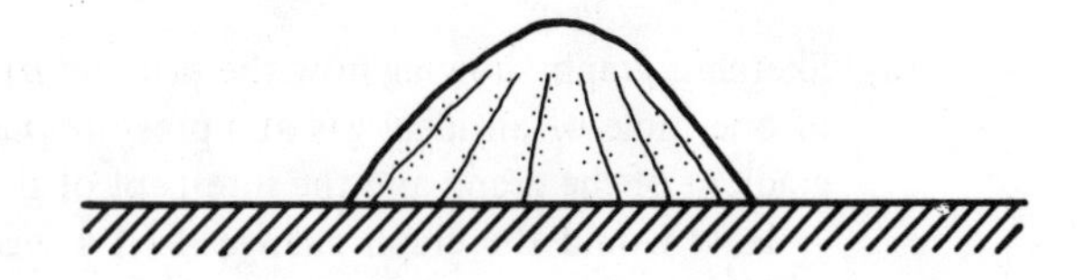

The diagram shows a shellfish called a limpet attached to a rock. When alarmed the creature reduces the pressure inside its rigid shell and sticks firmly to the rock. Assuming that the base of the shell is approximately circular and of effective diameter 30 mm, estimate the maximum force that could possibly be required to lift the limpet straight away from the rock. [2]

Solution 6.6

(a) (i) The molecules move at random.

They have negligible volume so most of the gas is empty space.

(ii) When smoke particles suspended in air are illuminated and viewed with a low-power microscope, they are seen to move randomly. This Brownian motion can only be because the gas molecules, which are invisible and bombarding the smoke particles, are themselves moving at random.

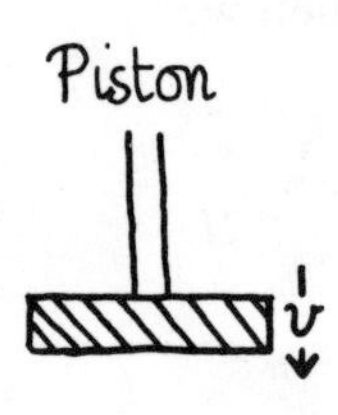

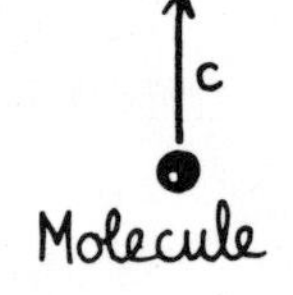

(b) The pressure increases because the molecules of the air now occupy a smaller volume. The number of collisions per second with a given area of the pump casing is increased and this greater rate of change of momentum means a greater force and so a greater pressure.

Molecules bouncing off the moving piston are speeded up because: The velocity of the molecule relative to the piston is $(c + v)$ towards it, and after an elastic bounce it is $(c + v)$ away from it.

Therefore the actual velocity of the molecule is $(c + v) + v = c + 2v$, which is faster than before.

The faster molecules will raise the average speed of the molecules of the air and hence its temperature.

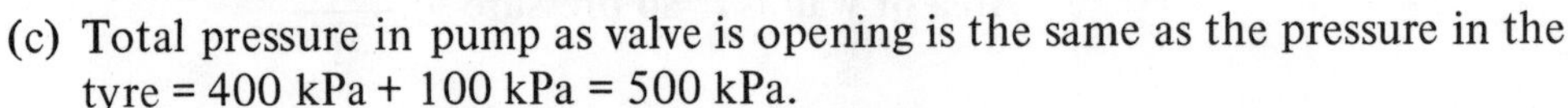

(c) Total pressure in pump as valve is opening is the same as the pressure in the tyre = 400 kPa + 100 kPa = 500 kPa.

Using the gas law $\dfrac{p_1 V_1}{T_1} = \dfrac{p_2 V_2}{T_2}$ and 27°C = 300 K

$$\Rightarrow \quad T_2 = \frac{T_1 p_2 V_2}{p_1 V_1} = \frac{(300 \text{ K})(500 \text{ kPa})(2.5 \times 10^{-5} \text{ m})}{(100 \text{ kPa})(1.0 \times 10^{-4} \text{ m})}$$

$$= 375 \text{ K, which is } 102°\text{C}$$

(d) $p = 1.0 \times 10^5 \text{ Pa} = 1.0 \times 10^5 \text{ N m}^{-2} = F/A$

The limpet covers an area $A = \dfrac{\pi d^2}{4} = \dfrac{\pi (0.030 \text{ m})^2}{4} = 7.1 \times 10^{-4} \text{ m}^2$

Be careful with diameters and radii. Notice the 1/4.

The push of the air on the limpet = pA
$= (1.0 \times 10^5 \text{ N m}^{-2})(7.1 \times 10^{-4} \text{ m}^2) = 71 \text{ N}$

Example 6.7

(a) Sketch a graph showing how the product pV_m varies with θ, where V_m is the volume of one mole of an ideal gas at a pressure p and a Celsius temperature θ. What are the gradient of the graph and the intercept of the graph on the temperature axis?

How would the graph change if (i) a second mole of the same gas were added to the first, (ii) the original gas were replaced by one mole of another ideal gas having twice the relative molecular mass of the first? [4]

(b) An ideal gas has a molar mass of 4.00 g mol^{-1}. The total kinetic energy of the molecules of a mass m of this gas is 375 J at a temperature of 27°C. Calculate (i) the kinetic energy of the molecules at a temperature of 127°C, (ii) the mass of the gas.

Molar gas constant = 8.32 J mol^{-1} K^{-1}.

Solution 6.7

(a) For an ideal gas $pV = nRT$, so a graph of pV against T is a straight line through the origin with gradient nR.

As $\theta/°C = T/K + 273$, the pV against θ graph has the same slope but looks as shown.

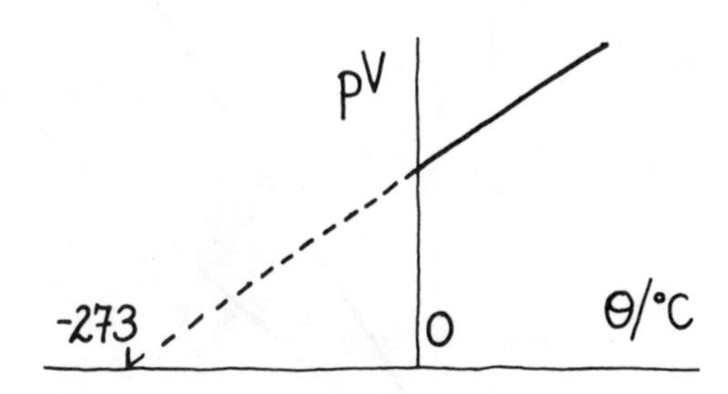

The gradient is (1 mol) $R = 8.32$ J K^{-1} and the intercept when pV is zero = -273°C.

The graph is drawn for pV and not for pV_m as in (ii) below you are asked what the graph would be like for 2 mol.
Generally, $pV = nRT$ is safer than $pV_m = RT$.

(i) The gradient would be (2 mol)R = 16.64 J K^{-1}, i.e. twice as steep, but still passing through -273°C.

(ii) There would be no change for it is still one mole and R is the same for any gas.

(b) Total (translational) kinetic energy = $N \times \frac{1}{2}m\overline{c^2}$, where there are N molecules each of mass m.
But $pV = nRT = \frac{1}{3}Nm\overline{c^2}$ so that $\frac{1}{2}Nm\overline{c^2} = \frac{3}{2}nRT$.

(i) The total kinetic energy is proportional to T so if T rises from (273 + 27) K to (273 + 127) K, that is from 300 K to 400 K, the total k.e. rises to $\frac{4}{3}$(375 J) = 500 J.

(ii) From 375 J = $\frac{3}{2}nRT$, $n = \dfrac{2\,(375\text{ J})}{3\,(8.32\text{ J mol}^{-1}\text{ K}^{-1})\,(300\text{ K})}$

$= 0.100$ mol

Mass of one mole is 4.00 g,
so mass of 0.100 mol = 0.40 g

Example 6.8

(a) The kinetic theory of gases predicts that the root mean square (r.m.s.) speed of the molecules of an ideal gas is given by the expression $(3p/\rho)^{\frac{1}{2}}$, where p is the pressure and ρ is the density of the gas.

The graphs show how the pressure of oxygen gas depends upon its density at two different constant temperatures, T and 300 K.

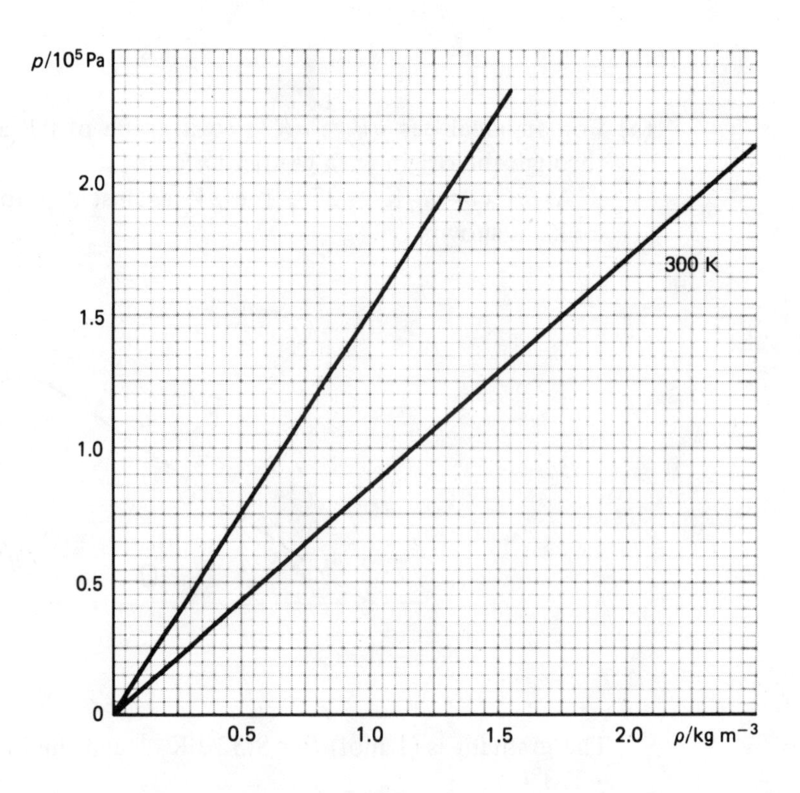

(i) Use the graph to calculate a value for the r.m.s. speed of the oxygen molecules at 300 K. Explain your working. [4]

(ii) Is the temperature T higher or lower than 300 K? Explain your reasoning. [2]

(iii) The graphs above are based upon experimental results. What conclusion can you draw from them about the behaviour of oxygen? [2]

(iv) Outline a simple experimental procedure for investigating how the pressure of a known mass of air varies as its density changes at room temperature. Include a labelled diagram of the apparatus used. [6]

(b) The graphs below show how the speeds of the molecules in an ideal gas are distributed at two temperatures. Use them to help you answer the following questions.

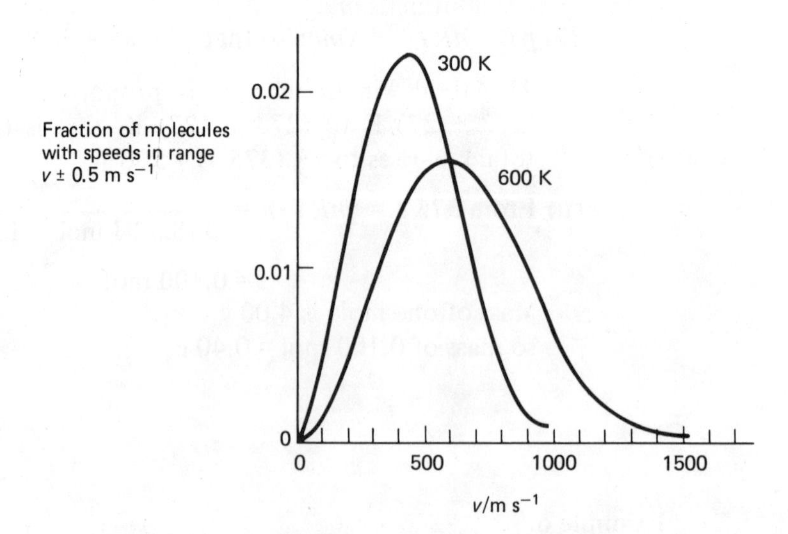

(i) In what two main ways does the temperature appear to affect the distribution of speeds? [2]

(ii) What is the value of v for which the fraction of molecules with speeds in the range $v \pm 0.5$ m s^{-1} is a maximum at a temperature of 300 K? How does this value compare with the r.m.s. speed calculated in (a) (i) above? [2]

(L)

Solution 6.8

(a) (i) $p = \frac{1}{3}\rho\overline{c^2}$, and at $\rho = 1.0$ kg m^{-3} the graph gives $p = 0.86 \times 10^5$ Pa.

$$\therefore \overline{c^2} = \frac{3p}{\rho} = \frac{3\ (0.86 \times 10^5\ \text{N m}^{-2})}{1.0\ \text{kg m}^{-3}} = 2.6 \times 10^5\ \text{m}^2\ \text{s}^{-2}$$

So $v_{\text{rms}} = 510$ m s^{-1}

(ii) $\overline{c^2}$ for temperature T is greater than for 300 K because p is greater at the same density. T must be greater than 300 K since $\overline{c^2}$ is proportional to T.

(iii) As the graphs are straight lines oxygen is behaving like the ideal gas used in the kinetic theory.

(iv) For a fixed mass of gas, density is inversely proportional to volume so we can do an experiment to study how the pressure of a fixed mass of gas varies with volume when the temperature is constant (Boyle's experiment).

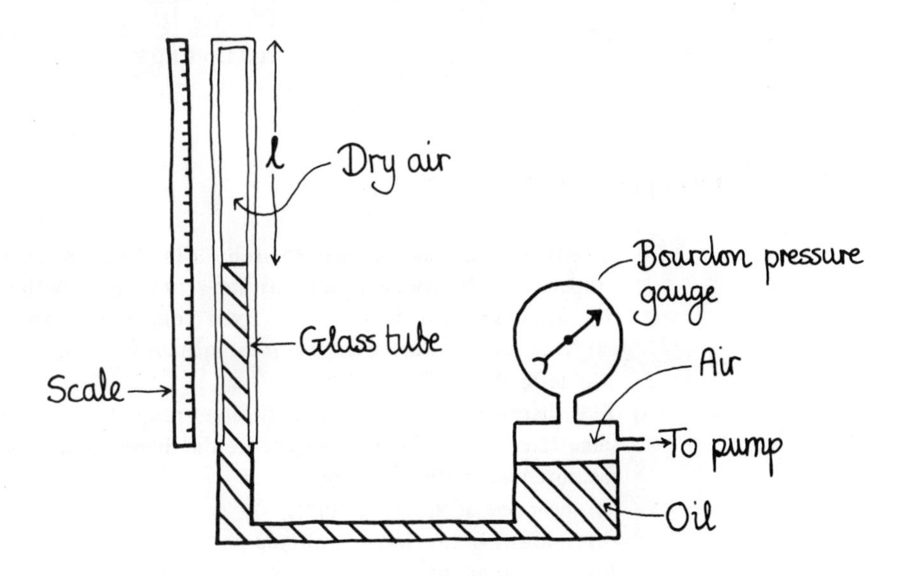

Increase pressure by about 20 kPa using the pump.

Wait to allow air to get back to room temperature.

Read steady value of the length of the air column l. This is proportional to the volume of the air.

Calculate $1/l$ and plot on a graph against pressure. $1/l$ is proportional to density.

Increase the pressure again repeating the procedure until the pressure is about 300 kPa or 3 atmospheres.

(b) (i) At the higher temperature the mean speed of the molecules is higher.

And there is a wider range of speeds and fewer molecules have speeds near the mean.

(ii) This speed is 420 m s^{-1} which is less than the r.m.s. speed of 510 m s^{-1}.

Average is not the same as the most popular.

Example 6.9

The average energy of an atom of a solid at a kelvin temperature T is $3kT$, where k is the Boltzmann constant. Calculate (a) the average energy of a copper atom in a mole of copper at a temperature of 300 K, and (b) the total energy of all the atoms in a mole of copper at the same temperature.

If this total energy could be transformed into linear kinetic energy of the whole mass of copper, at what speed would the mass be travelling?

The Boltzmann constant = 1.38×10^{-23} J K^{-1}.
The Avogadro constant = 6.02×10^{23} mol^{-1}.
Mass of 1 mole of copper atoms = 0.064 kg. [5]

(L)

Solution 6.9

(a) Average energy of an atom = $3kT$
$= 3\,(1.38 \times 10^{-23}\text{ J K}^{-1})\,(300\text{ K}) = 1.24 \times 10^{-20}\text{ J}$

(b) The Avogadro constant tells us the number of atoms in a mole so:
Energy per mole = $(6.02 \times 10^{23}\text{ mol}^{-1})\,(1.24 \times 10^{-20}\text{ J})$
$= 7.48 \times 10^{3}\text{ J mol}^{-1}$
Linear kinetic energy = $\frac{1}{2}$ (mass) (speed)^2

$$\text{so speed} = \sqrt{\frac{2 \times \text{k.e.}}{\text{mass}}} = \sqrt{\frac{2\,(7.48 \times 10^{3}\text{ J})}{(0.064\text{ kg})}} = 483\text{ m s}^{-1}$$

Example 6.10

(a) A temperature scale can be set up by using fixed points in conjunction with some property which varies continuously and reproducibly with temperature.
(i) What is meant by the term 'fixed points'? Give two examples of such points.
(ii) One such property is the resistance of a platinum coil. How is temperature defined on such a scale? [6]

(b) Two thermometers are based on different properties but they are calibrated using the same fixed points. To what extent are the thermometers likely to agree when used to measure temperature
(i) near one of the fixed points;
(ii) midway between the two fixed points?
Justify your answers. [5]

(AEB 1984)

Solution 6.10

(a) (i) Fixed points are temperatures at which certain changes take place under specified conditions.

Pure ice at atmospheric pressure melts at a temperature chosen to be 0° for the Celsius scale.

Pure water at standard atmospheric pressure boils at a temperature chosen to be 100°C.

(ii) Let R_T be the resistance of the coil at temperature T. R_0 and R_{100} are the resistances of the coil at the fixed points mentioned in (i).
T is defined by the equation:

$$T = \left(\frac{R_T - R_0}{R_{100} - R_0}\right)(100°\text{C})$$

(b) (i) Near a fixed point the temperatures will agree quite closely.
(ii) Midway any difference between the scales is likely to be near its greatest value.

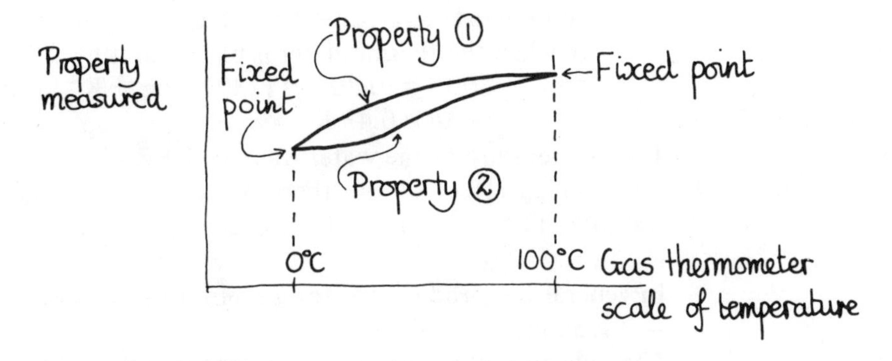

The graphs must both pass through the fixed points.
Neither property is necessarily linear when measured on a gas scale of temperature.

Neither scale is 'wrong', they are simply different.

Example 6.11

The resistance of the element of a platinum resistance thermometer is 2.00 Ω at the ice point and 2.73 Ω at the steam point. What temperature on the platinum resistance scale would correspond to a resistance value of 8.34 Ω? Measured on the gas scale, the same temperature corresponded to a value of 1020°C. Explain the discrepancy. [5]
(L)

Solution 6.11

If $R = 8.34\ \Omega$ then on the platinum resistance scale

$$\theta = \left(\frac{R_\theta - R_0}{R_{100} - R_0}\right)(100°\text{C}) = \left(\frac{8.34\ \Omega - 2.00\ \Omega}{2.73\ \Omega - 2.00\ \Omega}\right)(100°\text{C}) = 868°\text{C}$$

The gas scale and the platinum resistance scale must agree at 0°C and 100°C, but will not necessarily agree at other temperatures.

There is a difference, but it is not a discrepancy. In other words there is nothing wrong. You could look up tables showing you how to convert temperatures on the platinum scale to absolute gas temperatures.

Example 6.12

The element of a resistance thermometer is of mass 0.013 kg and has a specific heat capacity of 450 J $\text{kg}^{-1}\ \text{K}^{-1}$. Initially it is at room temperature, for which a reading of 17.1°C was obtained. It is then completely immersed in 0.30 kg of liquid of specific heat capacity 2300 J $\text{kg}^{-1}\ \text{K}^{-1}$ and gives an equilibrium reading of 68.4°C.

(a) What is the temperature of the liquid just before the thermometer was immersed? Neglect the heat capacity of the container. [4]

(b) How could the cooling effect of the thermometer be made less significant? [2]

Solution 6.12

(a) The energy transfer to the thermometer in warming from 17.1°C to 68.4°C, a rise of 51.3 K, is:
$\Delta Q = mc\Delta\theta = (0.013\ \text{kg})\ (450\ \text{J kg}^{-1}\ \text{K}^{-1})\ (51.3\ \text{K}) = 300\ \text{J}$

This energy must have come from the hot liquid so that it changes in temperature when the thermometer is placed in it by $\Delta\theta$, where

$$300 \text{ J} = (0.30 \text{ kg})(2300 \text{ J kg}^{-1} \text{ K}^{-1})\Delta\theta$$
$$\Rightarrow \Delta\theta = 0.43 \text{ K}$$

The temperature of the water was $(68.4 + 0.43)°\text{C} = 68.8°\text{C}$.

(b) The cooling effect of the thermometer could be made less significant by warming it under a hot tap before using it.

In general the product *mc* for a thermometer should be as small as possible – it is about 6 J K^{-1} here.

This question illustrates an important principle that taking a measurement can affect what is being measured. Another example is a low-resistance voltmeter, see Example 4.10.

Example 6.13

(a) (i) Explain the terms *specific heat capacity, specific latent heat*, and *internal energy*. [6]

(ii) Why is a distinction between the specific heat capacity at constant pressure and that at constant volume important for gases, but less important for solids and liquids? [3]

(b) Describe an accurate method to measure the specific latent heat capacity of vaporisation of a liquid boiling at atmospheric pressure. [10]

(c) Calculate the external work done and the internal energy gained when 1.0 kg of water at 100°C and 1.0×10^5 Pa pressure is converted to steam. Take the density of steam under these conditions to be 0.58 kg m^{-3}, the specific latent heat capacity of water to be 2.3×10^6 J kg^{-1}, and the density of water to be 1000 kg m^{-3}. [7]

(d) The specific latent heats of fusion of substances have markedly different values from those of their specific latents heats of vaporisation at the same pressure. Explain in general terms how these differences arise. [4]

(OLE)

Solution 6.13

(a) (i) Specific heat capacity (s.h.c.) is defined from the equation:

$$\Delta Q = mc\Delta T$$

where ΔQ is the energy supplied by heating, m the mass of the substance, c the s.h.c. and ΔT the change in temperature. The s.h.c. tells you how easy it is to raise the temperature of a substance.

The specific latent heat (s.l.h.) of a change of state is defined from the equation:

$$\Delta Q = ml$$

where ΔQ is the energy supplied by heating, m the mass and l the s.l.h.

There is no change of temperature during the change of state.

Internal energy is the energy a substance has because of the k.e. of the random motion of its molecules and the p.e. they have owing to their separation.

Internal energy (ΔU) is sometimes loosely called heat, but this introduces a confusion between it and the energy transferred by heating (ΔQ). Heating is simply the process by which energy is transferred from hot to cold.

The distinction is necessary to understand the first law of thermodynamics which states that the energy transferred by heating plus the energy supplied by doing mechanical work on the gas is equal to the increase in internal energy.

$$\Delta Q + \Delta W = \Delta U$$

(ii) The volumes of solids and liquids change very little when heated at constant pressure, compared with the volume changes for gases for the same temperature change.

Thus solids and liquids do very little mechanical work in expanding. The difference in energy is tiny between when they are allowed to expand and when they are not allowed to expand.

(b)

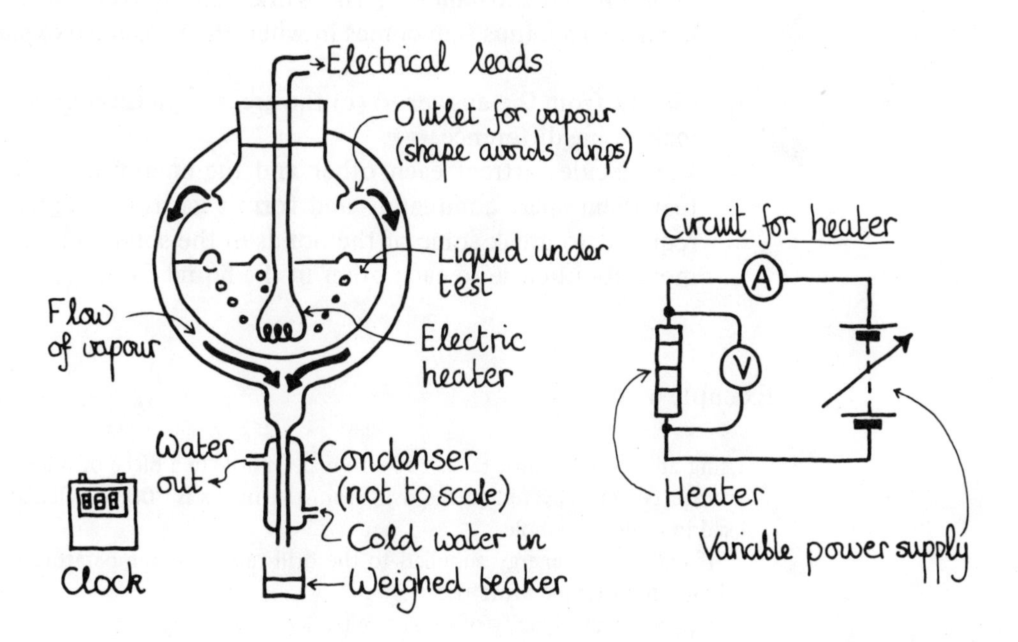

Once the liquid is boiling steadily and liquid has been dripping from the condenser for some time a previously weighed beaker is used to collect the liquid for a measured time.

Rate of boiling of liquid, $\frac{dm}{dt} = \frac{\text{change in mass of beaker}}{\text{time}}$

The voltmeter reading and ammeter reading are multiplied to give the electrical power.

Since the temperature of the boiling liquid is fixed, the rate of loss of energy will be about the same at two different powers. Call this power loss dH/dt.

The experiment is repeated boiling the liquid with a different power.

Power boiling away liquid = (electrical power) – (power loss)

$$\left(\frac{dm}{dt}\right)_1 l = V_1 I_1 - \frac{dH}{dt} \quad \text{and} \quad \left(\frac{dm}{dt}\right)_2 l = V_2 I_2 - \frac{dH}{dt}$$

There are two equations and two unknown quantities. It is thus possible to calculate the l and dH/dt.

Most of the information on the experiment is included in the diagram.

(c) The volume of 1.0 kg of steam = $\frac{m}{\rho} = \frac{1.0 \text{ kg}}{0.58 \text{ kg m}^{-3}} = 1.72 \text{ m}^3$

The volume of 1.0 kg of water = $\frac{1.0 \text{ kg}}{1000 \text{ kg m}^{-3}} = 0.001 \text{ m}^3$

The change in volume is therefore 1.72 m^3.

$\Delta W = -p\Delta V = -(1.0 \times 10^5 \text{ Pa})(1.72 \text{ m}^3) = -1.72 \times 10^5 \text{ J}$

The remainder of the energy supplied by heating increases the internal energy:

$\Delta U = \Delta Q + \Delta W = (2.3 \times 10^6 \text{ J}) + (-1.72 \times 10^5 \text{ J}) = 2.1 \times 10^6 \text{ J}$

ΔW can cause confusion. Be clear whether you have defined it as the work done *on* the substance or the work done *by* the substance. Here it is the former so a minus sign comes in when the substance expands.

(d) Clearly from the answer to (c) the great expansion on vaporisation does not make a big difference.

Molecules attract each other and the energy that has to be supplied to tear them apart completely and form a gas is much greater than the energy required to break some of the bonds in the solid and allow the molecules to move about close to each other in the liquid form.

Example 6.14

Using an electric drill it takes 250 s to make a hole in a piece of brass of mass 0.65 kg. If the average power delivered to the drill from the mains is 300 W, calculate how much energy is used in drilling the hole.

If 80% of the energy supplied to the drill raises the temperature of the brass, what is its initial rate of temperature rise?

Specific heat capacity of brass = 390 J kg^{-1} K^{-1}. [6]

Solution 6.14

300 W = 300 J s^{-1}

In 250 s the energy used = (300 J s^{-1}) (250 s) = 75 000 J

80% of 300 W = 240 W

Therefore in the first second 240 J go to warm the brass.

Using $\Delta Q = mc\Delta\theta$

$$240 \text{ J} = (0.65 \text{ kg})(390 \text{ J kg}^{-1} \text{ K}^{-1})\Delta\theta$$

$$\Rightarrow \Delta\theta = 0.95 \text{ K}$$

So the initial rate of rise of temperature is 0.95 K s^{-1}.

$\Delta Q = mc\Delta\theta$ can be rewritten as $\frac{dQ}{dt} = mc\frac{d\theta}{dt}$ if you want to go straight to $d\theta/dt$ and are happy with the notation. This gives a neater solution.

Example 6.15

(a) The specific heat capacities of air are 1040 J kg^{-1} K^{-1} measured at constant pressure and 740 J kg^{-1} K^{-1} measured at constant volume. Explain briefly why the values are different.

(b) A room of volume 180 m^3 contains air at a temperature of 16°C having a density of 1.13 kg m^{-3}. During the course of the day the temperature rises to 21°C. Calculate an approximate value for the amount of energy transferred to the air during the day. Assume that air can escape from the room, but no fresh air enters. Explain your reasoning. [5]

(L)

Solution 6.15

(a) To increase the temperature of a given mass of air at constant pressure or constant volume involves increasing the internal energy by the same amount ΔU.

But the specific heat capacity $c = \Delta Q/m\Delta\theta$ and though $\Delta Q = \Delta U$ at constant volume, at constant pressure $\Delta Q > \Delta U$ because the gas is then working as it expands $\Delta W = p\Delta V$ and this energy comes from ΔQ.

That is ΔQ for $c_p > \Delta Q$ for c_v and so $c_p > c_v$.

(b) Rise in temperature = 21 K − 16 K = 5 K, and this warming takes place at constant pressure assuming atmospheric pressure is constant.
Initial mass of air in room = $(1.13 \text{ kg m}^{-3})(180 \text{ m}^3) = 203$ kg
Some leaks out so assume mass warmed is 200 kg
From $\Delta Q = mc_p\Delta\theta$, the energy transferred to the air is

$$\Delta Q = (200 \text{ kg})(1040 \text{ J kg}^{-1} \text{ K}^{-1})(5 \text{ K}) = 1.04 \times 10^6 \text{ J}$$

that is about 1 MJ

In (a) you are using the first law of thermodynamics $\Delta U = \Delta Q + \Delta W$ where ΔU is the change in internal energy, ΔQ the energy supplied by heating and ΔW is the work done *on* the gas. As the gas does work here, ΔW in this equation will be negative.
Several assumptions have been made in (b). The mass of air leaking out of the room depends on the increase in the absolute temperature from 288 K to 294 K, an increase of less than 2%.

Example 6.16

(a) Legend has it that during Mr and Mrs J. P. Joule's honeymoon in Switzerland Mr Joule used a sensitive thermometer to investigate the difference in temperature between water at the top of a 50 m high waterfall and water at the bottom.

(i) Taking the specific heat capacity of water to be 4200 J kg^{-1} K^{-1} and assuming the potential energy of the water all went into heating it up, calculate the highest temperature difference he could have measured.

(ii) Explain what is meant by sensitive and how a mercury in glass thermometer could be made to be sensitive. [6]

(b)

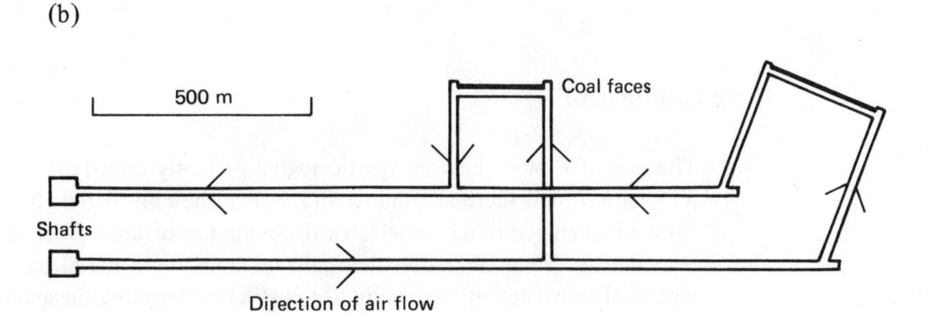

The diagram on the previous page shows the network of underground roadways in a coal mine. Ventilation is crucial in a mine to avoid the build up of dangerous gases such as methane and carbon monoxide. Fresh air is pumped down one shaft and stale air comes up the other shaft having travelled all round the mine. The 'wind' in one of the main roadways near the shafts is found to be of speed 5 m s^{-1} and the cross sectional area of the tunnel is 9 m^2.

(i) Estimate the mass of air being pumped into the mine each second.

(ii) The air also carries away the heat produced in the working mine. Estimate the temperature difference between the stale air and the fresh air if the total power of the lights, men and machines in the mine is 0.5 MW. [6]

(Take density of air as 1 kg m^{-3} and s.h.c. of air as 1 kJ kg^{-1} K^{-1}.)

Solution 6.16

(a) (i) For a mass of water m going over a waterfall the loss of gravitational energy (g.p.e.) = gain in internal energy (heat)

$$mg\Delta h = mc\Delta\theta$$

$$\therefore \Delta\theta = \frac{g\Delta h}{c} = \frac{(10 \text{ N kg}^{-1})(50 \text{ m})}{4.2 \times 10^3 \text{ J kg}^{-1} \text{ K}^{-1}} = 0.12 \text{ K}$$

If any of the g.p.e. lost is not used to heat the water the temperature rise will be less than this; 0.12 K is a maximum.

(ii) A sensitive thermometer is one on which you can read very small temperature changes.

A sensitive thermometer may not be accurate, e.g. it may read 17.42°C when the actual temperature is 17.98°C.

For a mercury in glass thermometer a very narrow bore and a large mercury reservoir are needed.

(b) (i) Rate of flow of air $= \dfrac{dm}{dt}$ = density × speed × area

$= (1 \text{ kg m}^{-3})(5 \text{ m s}^{-1})(9 \text{ m}^2)$

$= 45 \text{ kg s}^{-1}$ (5×10^1 kg s^{-1} to one sig. fig.)

(ii) Power $= \dfrac{dm}{dt} \times c\Delta\theta$

$$\text{so } \Delta\theta = \frac{(5 \times 10^5 \text{ W})}{(45 \text{ kg s}^{-1})(1000 \text{ J kg}^{-1} \text{ K}^{-1})} = 11 \text{ K} \quad (10 \text{ K to 1 sig. fig.})$$

There are many complications with the air flow and so the data are only given to 1 sig. fig. You should therefore only give the final answer to 1 sig. fig.

Example 6.17

The rate of flow of heat energy through a perfectly lagged metal bar of area of cross section X, length d, and thermal conductivity k, may be considered to be analogous to the rate of flow of charge through an electrical conductor of area of cross section A, length l, and resistivity ρ. Show that the 'thermal resistance' of a metal bar which corresponds to the electrical resistance of the conductor is d/kX. Extending the analogy to thermal conductors in series show that the effective thermal conductivity K of a composite wall consisting of two parallel sided layers of materials of thickness d_1 and d_2 and thermal conductivities k_1 and k_2 respectively is given by the expression

$$K = \frac{d_1 + d_2}{d_1/k_1 + d_2/k_2}$$

[8]
(AEB 1983)

Solution 6.17

$$\text{Rate of heat flow through bar} = \frac{kX(\theta_1 - \theta_2)}{d}$$

where $(\theta_1 - \theta_2)$ is the temperature difference between the ends of the bar. This is analogous to the potential difference V in the electrical case.

The Ohm's law formula is $I = V/R$

$1/R$ is the electrical conductance and so is analogous to kX/d.

The resistance R is analogous to d/kX the 'thermal resistance'.

'Thermal resistances' will add up when in series as do electrical resistances. So the thermal resistance of the whole wall $= (d_1 + d_2)/KX$

$$= d_1/k_1X + d_2/k_2X$$

Multiplying through by X and rearranging we get the equation required:

$$K = \frac{(d_1 + d_2)}{d_1/k_1 + d_2/k_2}$$

The analogy between heat flow and charge flow is a useful one. Problems in heat can be solved using very familiar results from circuit theory. It is often only a matter of habit that people use resistance and resistivity in electrical problems and conductance and conductivity in heat flow problems.

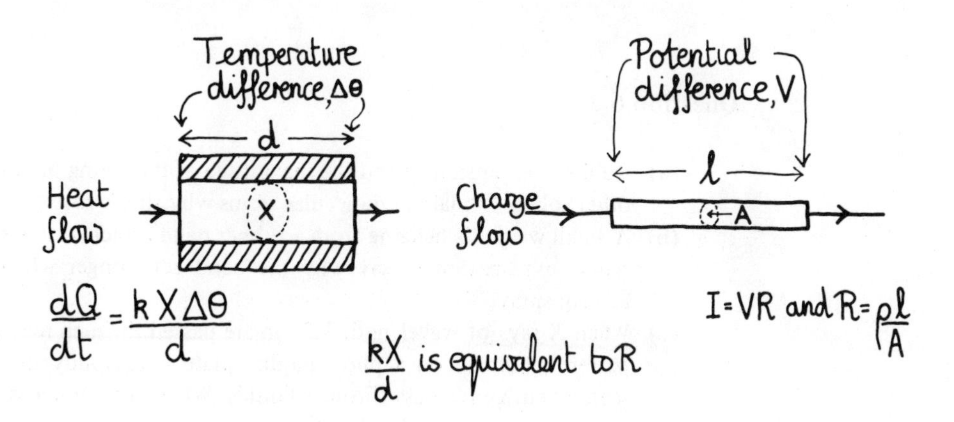

6.7 Questions

Question 6.1

(a) What is meant by (i) elastic behaviour, (ii) plastic behaviour of a wire when it is stretched? [2]

(b) (i) Describe how you would investigate the elastic and plastic properties of a soft copper wire under increasing tensile loads up to its breaking point. [6]

(ii) Draw a graph of the results you would expect to obtain, and label its principal features. [4]

(iii) Interpret the graph in terms of the forces between atoms in the wire and of their arrangement in the wire. [4]

(c) A lift in a skyscraper has a total mass of 8000 kg when loaded. It is hung from light cables made of steel of breaking stress 0.50×10^9 N m^{-2}. These cables will support a static load of 72 000 kg before they break.

Take the Young modulus for steel to be 2.0×10^{11} Pa.

Calculate:

(i) the total cross-sectional area of the lift cables; [3]

(ii) the static extension of the cables when the lift is at rest at ground-floor level, if the height to the winding gear at the top of the building is 350 m; [3]

(iii) the elastic strain energy stored in the cables when the lift is at rest at ground-floor level. [3]

The lift now ascends at a steady speed of 8.0 m s^{-1}.

(iv) Calculate the power needed to raise the lift.

(v) How could the lift system be designed to reduce significantly this large power requirement? [3]

(OLE)

Question 6.2

The lift in a tall building has a mass of 800 kg when full and is supported on cables with a breaking stress of 5.0×10^8 Pa.

(a) Draw diagrams showing the forces on the lift when it is

(i) going up with an upward acceleration of 2 m s^{-2}

(ii) going up at a steady speed of 6 m s^{-1}

(iii) going down with a downward acceleration of 4 m s^{-2}.

Label the forces with their magnitudes.

(b) The cables must never reach 10% of the breaking stress and the lift must be able to accelerate upwards at 3 m s^{-2}. What is the minimum cross sectional area of the cables? [8]

Question 6.3

(a) Rubber's extensibility and the magnitude of its Young modulus differ markedly from other solids. Explain in molecular terms why this is so. [3]

(b) A small weight is hanging from a rubber band attached to a retort stand. Use a microscopic explanation to say why the band gets longer when rapidly cooled with an aerosol spray. [2]

(c) When X-rays of wavelength 1.5 nm are passed through relaxed rubber the diffraction pattern observed on a photographic plate is a cloudy disc similar to that which is seen when X-rays pass through liquids. When the rubber is stretched dots appear in the pattern. Explain simply why this is so. [2]

Question 6.4

(a) Explain what is meant by an ideal gas. State two respects in which a real gas differs from an ideal gas. Under what conditions do most gases approach the ideal closely enough for many purposes? [5]

(b) Using a simple kinetic theory it may be shown that, for an ideal gas, $p = \frac{1}{3}\rho\overline{c^2}$, where p is the pressure of the gas, ρ its density and $\overline{c^2}$ the mean square speed of the molecules. Use this formula to calculate the root mean square speed of hydrogen molecules at s.t.p. if standard pressure is 1.01×10^5 Pa and the density of hydrogen at s.t.p. is 0.090 kg m^{-3}. Sketch graphs, on the same axes, to indicate how the molecular speeds are distributed at two different temperatures. State clearly the meanings of the quantities to which the axes of your graph correspond. Mark which curve corresponds to the higher temperature. [8]

(c) Show how the equation $p = \frac{1}{3}\rho c^2$ may be related to the equation of state for an ideal gas, $pV = nRT$, where p is the pressure, V the volume of n moles, R the molar gas constant and T the absolute temperature. Justify any assumption which you have to make. [5]

(d) The mass of one mole of oxygen molecules is 32.0 g and their root mean square speed is 460 m s^{-1} at 0°C. If the Avogadro number is 6.02×10^{23} mol^{-1}, obtain a value for the Boltzmann constant. [3]

(AEB 1983)

Question 6.5

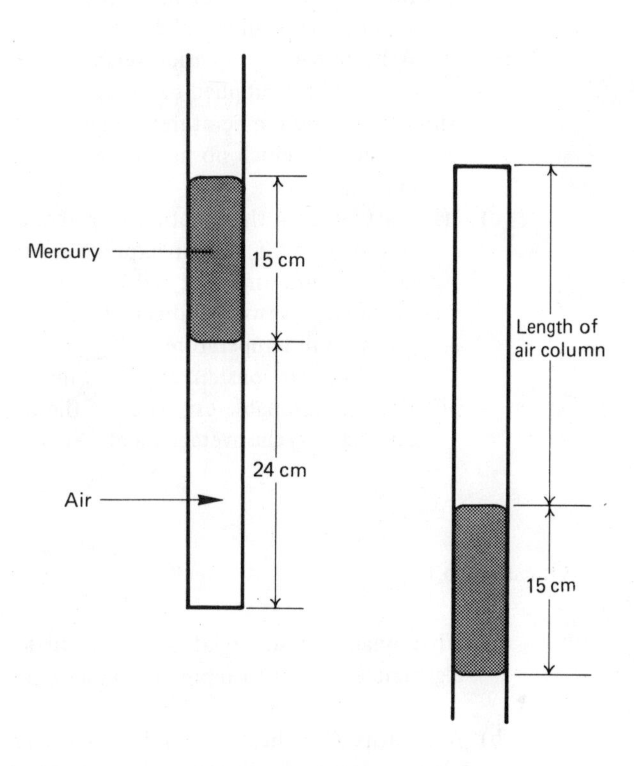

The figures show a fixed mass of air trapped by a column of mercury in a vertical tube. Atmospheric pressure is 75 cm of mercury. What will be the length, in cm, of the air column when the tube is inverted? Assume the temperature to be the same in both cases.

A 16 **B** 18 **C** 32 **D** 36 **E** 39

(AEB 1983)

Question 6.6

(a) The temperature of a hot liquid, measured on the empirical centigrade scale of a certain metallic resistance thermometer, is 78.7°.

(i) Why is this temperature scale described as *empirical* and *centigrade*?

(ii) The resistance of the thermometer at the ice and steam points is 12.35 Ω and 17.29 Ω respectively. Draw a graph to show how the resistance R of the thermometer depends on the empirical temperature t.

Read off from your graph the resistance of the thermometer when the temperature of the element is 78.7°. [7]

(b) The element of the thermometer in (a) has mass 0.015 kg and constant specific heat capacity 1.4×10^2 J kg^{-1} K^{-1}. Initially, it was in equilibrium at room temperature, for which a reading of 16.2° was obtained. It was then completely immersed in

0.19 kg of liquid of constant specific heat capacity 1.7×10^3 J kg^{-1} K^{-1}. The equilibrium temperature was 78.7°.

(i) Neglecting the heat capacity of the container, calculate the temperature of the liquid just before the thermometer was immersed.

(ii) Suggest ONE way in which the cooling effect of the element could be made less significant. [5]

(NISEC)

Question 6.7

(a) (i) Explain what is meant by the *internal energy* of a system.

(ii) Identify the forms of internal energy that are possessed by an ideal monatomic gas and a crystalline solid. [3]

(b) (i) Write down the equation relating ΔU, the change in internal energy of a system, ΔQ, the heat supplied to the system, and ΔW, the work done on the system.

(ii) Explain the temperature change that results when an ideal gas undergoes an expansion in which no heat enters or leaves the system, and in doing so performs external work. [3]

(c) (i) Starting from the kinetic theory equation $p = \frac{1}{3} nm \langle c^2 \rangle$ and the ideal gas equation $pV_m = RT$, find an expression for the internal energy U_m of one mole of an ideal gas, in terms of R and T.

(ii) Calculate, to two significant figures, the internal energy of one mole of an ideal gas at room temperature.

(Molar gas constant $R = 8.3$ J mol^{-1} K^{-1}.)

(iii) Making reasonable estimates of the quantities involved, calculate (to two significant figures) the average kinetic energy of a sprinter in a 100 m race. [6]

(NISEC)

Question 6.8

(a) What quantities are related by the first law of thermodynamics? Write down an algebraic equation to represent the law, and identify and explain each of the terms. [5]

(b) A quantity Q of heat is supplied to a sample of an ideal monatomic gas under reversible conditions. Explain how the first law of thermodynamics can be used to describe the changes that occur if the gas is maintained

(i) at constant volume; [2]

(ii) at constant pressure. [3]

State and explain whether each of these processes could be described as an isothermal process. [4]

(OLE)

6.8 Answers to Questions

6.1 (a) See a textbook.

(b) (ii) Up to the elastic limit the atoms return to their original position after the stress is removed. See section 6.3.

(c) (i) The weight of the 72 000 kg load is required, not its mass.

Answer 1.4×10^{-3} m^2

(ii) Use the stress when a force of 80 000 N is applied and not the breaking stress. From the strain you can calculate the extension.

Answer 0.10 m

(iii) The easiest method is to imagine a force gradually increasing to 80 000 N and finally extending the wire by 0.10 m. Then consider the average force during the increase.
Answer 4000 J

(iv) Since energy = average force × displacement
power = average force × velocity
Answer 640 kW

(v) The method should avoid wasting the gravitational potential energy lost as the lift descends.

6.2 (a) There are two forces in each case and you need to think about Newton's first and second laws.
(b) The area is 2.1×10^{-4} m^2.

6.3 (a) Questions do not always specify explanations in terms of microscopic structures, but such explanations are almost always necessary in questions on the behaviour of materials. The key to the explanation is the long chain molecules of which rubber is made.
(b) The long chain molecules are easier to stretch when cooler. Suggest why.
(c) The structure in diffraction patterns relies on an organised structure within the material.

6.4 (a) The explanation here is equivalent to listing the assumptions about molecules used in the kinetic theory. For the differences choose two assumptions which are not *exactly* true for a real gas like oxygen.
(b) $\overline{c^2} = \langle c^2 \rangle$. You should get c_{rms} = 1800 m s^{-1} to 2 sig. fig. For the graphs see worked example 6.8 on page 158. Be sure you understand the way in which the speed axis is labelled.
(c) Take $p = \frac{1}{3}\rho \langle c^2 \rangle$ and replace the density ρ by mass/volume, Nm/v where there are N molecules each of mass m.
Write an expression for pV in terms of $\langle c^2 \rangle$.
In $pV = nRT$ replace n by N/N_A.
Equate the two pV's and you should get, after simplifying,

$$\tfrac{1}{3}m \langle c^2 \rangle = \frac{R}{N_A} T \quad (\text{or } kT)$$

So assuming that the k.e. of a molecule $\frac{1}{2}m \langle c^2 \rangle$ is proportional to the absolute temperature T, the equations are related.
(d) In (c) you have a relation between m, $\langle c^2 \rangle$, k and T which will give $k = 1.37 \times 10^{-23}$ J K^{-1}.

Beware *m*, the mass of one molecule – it must be in kilograms when used in this relation.

6.5 In one case the pressure is higher than atmospheric pressure by 15 cm Hg and in the other case lower.
Answer D

6.6 (a) (i) Centigrade simply means a scale broken into 100 parts.

Often the term centigrade is loosely used to mean Celsius.

(ii) You should get 16.24 Ω from a graph like Fig. 6.3 on page 147. See worked example 6.11.

(b) Energy is conserved – see worked example 6.12.

6.7 (a) (i) You should refer to p.e. and k.e.

(b) (ii) The gas cools down – see worked example 6.6 for a similar discussion when a gas warms up.

(c) (i) Here n means the number density of molecules, i.e. the number per unit volume $n = N/V$. If $N = N_A$, the Avogadro number, then $V = V_m$. Using this plus the ideal gas equation given you should be able to get $U_m = \frac{3}{2}RT$.

(ii) At $\theta = 20°C$, $U_m = 3.6$ kJ.

(iii) You might get from 1000 J to 5000 J.

Some boards require estimates to be given to only one significant figure. Check what it says at the top of the paper.

6.8 (a) See worked example 6.13 for an explanation of the distinction between energy supplied by heating and the internal energy.

(b) At constant volume $\Delta W = 0$.
At constant pressure the gas does work against the external pressure.

(i) Cannot possibly be isothermal if $Q > 0$.

(ii) Cannot possibly be isothermal since $c_p = 0$.

7 Atomic and Nuclear Physics

7.1 You Should Recognise

Quantity	*Symbol*	*Unit*	*Comment*
electronic charge	e	coulomb, C	a constant = 1.60×10^{-19} C
electron mass	m_e	kilogram, kg	a constant = 9.11×10^{-31} kg
electric field strength	E	newton per coulomb, N C^{-1}	or V m^{-1}
magnetic flux density	B	tesla, T	
energy	E	joule, J	or electronvolt, 1 eV = 1.6×10^{-19} J
work function	Φ	volt, V	1 V = 1 J C^{-1}
the Planck constant	h	joule second, J s	a constant = 6.63×10^{-34} J s
wavelength	λ	metre, m	usually nm for e.m. waves
speed of electromagnetic waves *in vacuo*	c	metre per second, m s^{-1}	a constant = 3.00×10^8 m s^{-1}
momentum	p	newton second, N s	1 N s = 1 kg m s^{-1}
proton number	Z	(no unit)	atomic number
nucleon number	A	(no unit)	mass number
neutron number	N	(no unit)	
activity of a source	A	bequerel, Bq	1 Bq = 1 s^{-1}
detector count rate	A_d	per second, s^{-1}	or min^{-1}
radioactive half-life	$t_{1/2}$	second, s	may be many years
radioactive decay constant	λ	per second, s^{-1}	
unified atomic mass constant	u	kilogram, kg	1 u = 1.66×10^{-27} kg

Not listed are quantities, such as mass, length and time, which occur in all chapters and which are given in the list on page 7.

7.2 You Should be Able to Use

- Free electrons:
  electron accelerated through a p.d. V has k.e. $\frac{1}{2}mv^2 = eV$
  forces on electrons in electric and magnetic fields
  $$F_e = eE, \qquad F_m = Bev \quad (B \text{ perpendicular to } v)$$
  de Broglie's relationship $p = \dfrac{h}{\lambda}$
- Photons:
  photon energy $E = hf = \dfrac{hc}{\lambda}$

Einstein's photoelectric equation $hf = \frac{1}{2}mv_{max}^2 + e\Phi$
energy transitions in atoms $E_1 - E_2 \rightleftharpoons hf$

- Radioactive decay:
the decay law $\frac{dN}{dt} = -\lambda N$ or $N = N_0 e^{-\lambda t}$
half-life and decay constant $\lambda t_{1/2} = \ln 2 = 0.693$

- The nuclear atom:
for nucleons $A = Z + N$
notations for particles:
α-particle ^4_2He, β^--particle $_{-1}^{0}e$, β^+-particle ^0_1e (positron), neutron ^1_0n
notation for nuclei:
^A_ZX e.g. ^1_1H (proton), $^{14}_6C$, $^{222}_{86}Rn$
mass–energy conservation $E = mc^2$

7.3 Electron Physics

Millikan's experiment measures the charge on oil drops and shows that the charge is always $\pm Ne$ where N is a whole number, i.e. the charge is quantised.

By measuring the specific charge, e/m_e, of electrons, e.g. by studying the path of a beam of electrons in known electric and magnetic fields, the mass of an electron can be found. Similarly the isotopic composition of an element can be discovered by studying beams of ions.

Electrons accelerated through a potential difference V have energy eV and can excite or ionise atoms by colliding with them. A conduction electron in a metal needs to be given energy $e\Phi$, where Φ is the work function of the metal, to escape from its surface.

7.4 Photons and Energy Levels

Photons are packets of light energy (strictly electromagnetic wave energy). The energy is said to be quantised.

When light of one wavelength strikes a metal surface all the energy of each photon is given to one electron. In this photoelectric effect (i) increasing the intensity of the light increases the number of emitted electrons but not their energy, and (ii) increasing the frequency ($f \propto 1/\lambda$) of the light increases the energy of the emitted photons. An experiment to investigate (i) and (ii) is given in worked example 7.3 on page 186.

Atoms emit and absorb energy only in certain definite amounts. The emissions

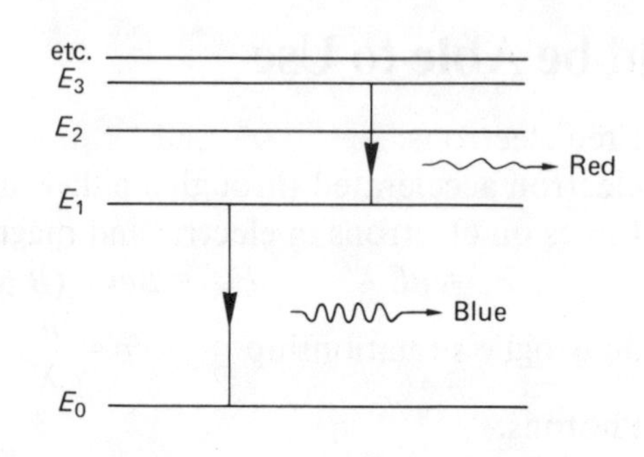

Figure 7.1

from gas atoms form line spectra, each line (colour) containing photons of one energy (and wavelength). Figure 7.1 is an energy level diagram for an atom; a transition from E_3 to E_1 may produce a 'red' photon, from E_1 to E_0 a 'blue' photon, etc. The squiggles representing the photons are only to help your understanding – we don't think they really look like this.

7.5 The Nuclear Atom and Radioactivity

Atoms consist of a very small positively charged nucleus made up of protons and neutrons. Almost all the mass of an atom is in its nucleus.

α-Particles are the same as high-speed helium nuclei; β^--particles are very high-speed electrons; γ-rays are high-frequency photons. Each is detected by its ability to cause ionisation, e.g. in the gas of a cloud chamber or GM tube.

Nuclear transformations are represented by equations of the form

(a) natural radioactivity $\quad {}^{222}_{86}\text{Rn} \rightarrow {}^{218}_{84}\text{Po} + {}^{4}_{2}\text{He}$

(b) artificial transmutation $\quad {}^{14}_{7}\text{N} + {}^{4}_{2}\text{He} \rightarrow {}^{17}_{8}\text{O} + {}^{1}_{1}\text{H}$

You do not need to remember the details of any of these equations.

Radioactive decay is a random phenomenon, but because so many nuclei are involved we can generally be sure of the result. A graph of the number, N, of the nuclei of a radioactive material against time is exponential. So is the graph of its activity $\mathrm{d}N/\mathrm{d}t$ against time, providing any background count has been allowed for. A graph of $\ln N$ or $\ln A$ against t is a straight line.

In a series of natural radioactive decays, starting from a long-lived isotope such as ${}^{238}_{92}\text{U}$, a state of radioactive equilibrium is reached. The activity of each isotope in the series is then the same.

Mass–energy is conserved in nuclear reactions. An individual disintegration results in an appreciable reduction in the total rest mass of the nuclei and particles involved, thus providing the k.e. of the emitted particles which will be of the order of MeV ($1\ \text{MeV} \equiv 1.8 \times 10^{-30}$ kg).

7.6 Wave–Particle Duality

Light (e.m. waves) exhibits both wave-like and particle-like properties. At a simple level light is transmitted as waves but interacts with matter as photons (particles).

Electrons (small particles) exhibit both particle-like and wave-like properties. At a simple level electrons obey Newton's laws of motion but they also diffract and superpose (waves).

7.7 Worked Examples

Example 7.1

A singly-ionised helium atom (${}^{4}_{2}\text{He}$) and a hydrogen (${}^{1}_{1}\text{H}$) ion are accelerated from rest through a potential difference V. The ratio of the final speed of the helium ion to that of the hydrogen ion is

A 1:4 **B** 1:2 **C** 1:1 **D** 2:1 **E** 4:1

(NISEC)

Solution 7.1

The He ion and the H ion have the same charge. Since they pass across the same potential difference they will gain the same kinetic energy. Since the He ion has four times the mass, v^2 will be four times smaller. Thus v will be half the value for the H ion. The ratio is 1:2.
Answer **B**

It is easy to be confused by 4 times smaller meaning the same as 1/4 as big.

Example 7.2

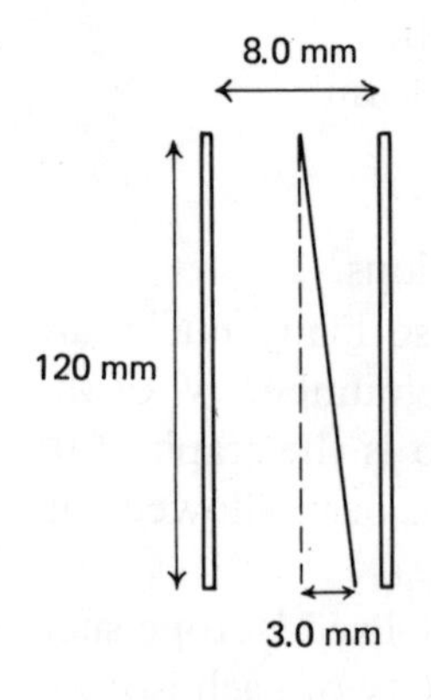

A charged water drop of weight 6.0×10^{-14} N falls at a terminal speed of 0.15 mm s^{-1} in air between two parallel plates, 120 mm long and placed 8.0 mm apart. When a p.d. of 15 V is applied between the plates the path of the drop is as shown in the diagram.

(a) Draw a diagram showing the forces acting on the drop. Explain how each of the forces arises. Calculate
  (i) the electrical force on the drop, and
  (ii) the charge on the drop, in terms of electron charges.

(b) What is the largest possible p.d. between the plates if the drop is not to hit the right hand plate?

Solution 7.2

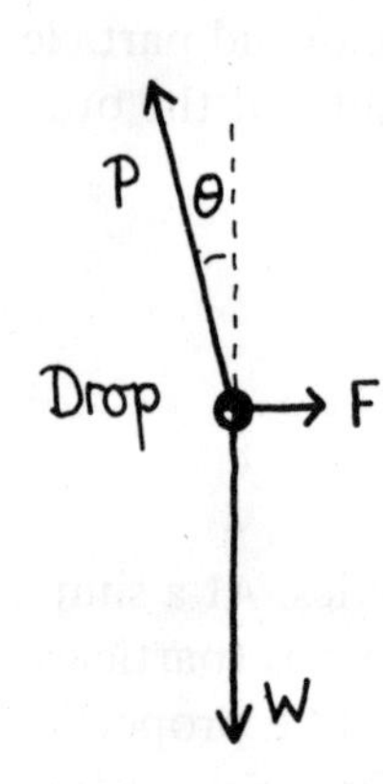

(a) W is the gravitational pull of the Earth on the drop.
F is the electrical force of the plates on the drop.
P is the viscous or frictional push of the air on the drop.

(b) (i) As the drop is moving at a constant speed it is in equilibrium, i.e.

$$P \sin \theta = F$$
$$P \cos \theta = W$$
$$\Rightarrow \frac{F}{W} = \tan \theta = \frac{3 \text{ mm}}{120 \text{ mm}}$$
$$\therefore \quad F = (6.0 \times 10^{-14} \text{ N}) \times \frac{3}{120} = 1.5 \times 10^{-15} \text{ N}$$

(ii) This force is produced by an electric field E where

$$E = \frac{V}{d} = \frac{15 \text{ V}}{8.0 \times 10^{-3} \text{ m}} = 1.88 \times 10^3 \frac{\text{V}}{\text{m}} \text{ or } 1.88 \times 10^3 \frac{\text{N}}{\text{C}}$$

$$F = (Ne)E, \text{ so } \quad Ne = \frac{F}{E} = \frac{1.5 \times 10^{-15} \text{ N}}{1.88 \times 10^3 \text{ NC}^{-1}} = 8.0 \times 10^{-19} \text{ C}$$

As $e = 1.6 \times 10^{-19}$ C, the drop has a charge of $5e$.

(c) A p.d. which is 4/3 times as big would increase F by 4/3 and move the drop over 4.0 mm as it fell.
The maximum p.d. = (15 V) $\times \frac{4}{3}$ = 20 V

Note that the speed of the drop does not enter the calculation. You must not assume that all the information given must be used. It is only useful here because it tells you that the drop is moving very slowly.

Example 7.3

A very tiny charged plastic ball is being held stationary between two horizontal charged plates. The p.d. between the plates is at 800 V to start with, but can be adjusted.

The supply to the plates is then switched off for a moment and a cover removed from a small radioactive source which is also between the plates. The cover is then replaced quickly and the supply switched back on. It is discovered that a p.d. of 267 V is required to keep the ball stationary.

The process is repeated many times and the various p.d.'s are recorded below. A minus sign indicates the p.d. was reversed. The figures are quoted in volts to the nearest volt.

800	267	160	−1600	−320	−533
800	400	320	229	229	200
800	1600	800	1600	−1600	−800
−533	−400	−400	−267	−178	−160
−133	−133	−100	−107	−178	−320
not held	−1600	1600	not held	800	1600

'Not held' indicates that however large the p.d. the ball could not be held steady.

(a) Explain why the radioactive source causes the charge on the ball to change. [2]

(b) What would happen if the supply were not switched off while the source was uncovered? [2]

(c) Why do the figures suggest that charge is quantised? Include in your explanation the theory of why the ball can be stationary. [4]

(d) Sometimes however big the p.d. between the plates the ball could not be held stationary and so it was exposed again to the source. Explain why it could not be held stationary. [2]

(e) The greatest p.d. that held the ball stationary was 1600 V. How many quanta of charge were on the ball when the required p.d. was (i) 800 V, (ii) 107 V? [4]

(f) The distance between the plates was measured as 4.0 mm and the diameter of the ball was found, by inspecting it under a powerful microscope, to be 2.9 μm. The density of the plastic of which the balls were made was 500 kg m^{-3}.

(i) What is the weight of the ball?

(ii) Use the results above to calculate a value for the charge on the electron. [6]

Solution 7.3

(a) The radiation from the source ionises the air and these ions come into contact with the ball and change its charge.

The more ionising radiations are alpha and beta so the source must radiate one of these.

(b) All the ions would be swept towards the charged plates without affecting the ball.

(c) Only certain values of p.d. are ever obtained suggesting the charge can only have certain values.

The weight of the ball $= mg = qE = qV/d$

E is the electric field, q the charge on the ball, V the p.d. and d the distance between the plates.

Since mg and d are constant during the experiment, qV must also be constant. Thus if V takes fixed values then q also takes only certain values.

(d) The ball must be uncharged so there is no upward force on it to balance the Earth's downward pull on the ball.

(e) 1600 V is the highest p.d. and so corresponds to the lowest charge.

(i) That suggests that 800 V corresponds to 2 units of charge because only half the field is required to produce the same force.

(ii) 1600 V/107 V = 14.95. This suggests that the charge is now 15 times the smallest possible charge.

(f) (i) Volume of ball = $\frac{4\pi}{3}\left(\frac{2.9 \times 10^{-6}\text{ m}}{2}\right)^3 = 1.28 \times 10^{-17}\text{ m}^3$

Weight = $mg = (500\text{ kg m}^{-3})(1.28 \times 10^{-17}\text{ m}^3)(10\text{ N kg}^{-1})$
$= 6.4 \times 10^{-14}\text{ N}$

(ii) Electric field when p.d. is 1600 V = 1600 V/0.004 m
$= 4.0 \times 10^5\text{ V m}^{-1}$

The charge is e at this p.d. so
$(4.0 \times 10^5\text{ V m}^{-1})e = 6.4 \times 10^{-14}\text{ N}$

$$e = \frac{6.4 \times 10^{-14}\text{ N}}{4.0 \times 10^5\text{ N C}^{-1}} = 1.6 \times 10^{-19}\text{ C}$$

Notice that remembering that E has units V m^{-1} and N C^{-1} can help you check your working. Also having a rough idea about the fundamental constants is helpful in questions where you calculate one from experimental results.

Example 7.4

(a) A beam of electrons, having a common velocity, enters a uniform magnetic field in a direction normal to the field. Describe and explain the subsequent path of the electrons. Explain whether a similar path would be followed if a uniform electric field were substituted for the magnetic field. [5]

(b) Describe an experiment for measuring the ratio of the charge to the mass of an electron. Show how the result is obtained from the measurements taken. [11]

(c) Electrodes are mounted at opposite ends of a low pressure discharge tube and a potential difference of 1.20 kV applied between them. Assuming that electrons are accelerated from rest, calculate the maximum velocity which they could acquire. Suggest a reason why the actual maximum velocity is likely to be less than the theoretical value. (Specific electron charge = -1.76×10^{11} C kg^{-1}.) [5]

(AEB 1983)

Solution 7.4

(a) When charged particles are moving at right angles to a magnetic field they experience a force at right angles to their motion and at right angles to the field. The 'sideways' force causes them to move along the arc of a circle. Their speed does not change, only the direction of their motion. The size of the force is given by $F = Bqv$.

In an electric field the charged particles experience a force in the direction of the field if they are positive or in the opposite direction if they are negative. If the field is uniform the force will always be in the same direction and so the electrons will move along parabolic paths and not circular ones. Their paths are like those of projectiles moving in the uniform gravitational field close to the surface of the Earth.

You may prefer to answer here with two sketches showing (i) the *B*-field into the page and a circular path and (ii) the *E*-field parallel to the page and a parabolic path. The forces could be marked on the sketches with force arrows and their sizes, *Bev* and *eE*, given.

(b)

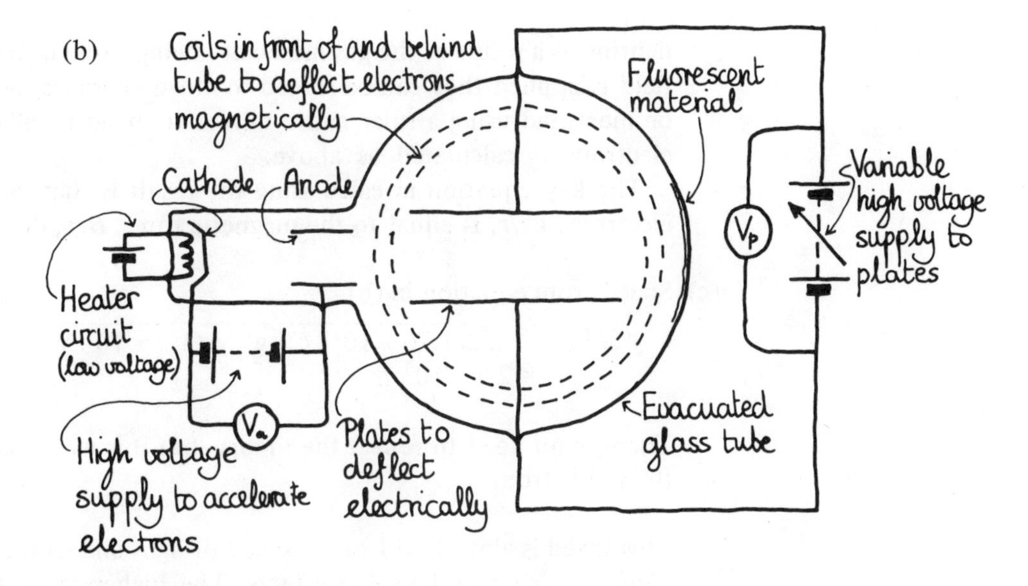

A labelled diagram here is vital.

Thermionic electrons from the glowing filament are accelerated across a p.d. of V_a. Their k.e. is $\frac{1}{2}mv^2 = eV_a$, so their speed is $v = \sqrt{2eV_a/m}$.

They pass into a region where two charged parallel plates can provide a uniform electric field whose strength and direction can be varied by adjusting V_p. We know $E = V_p/d$ where d is the distance between the plates. Also in this region two identical coils placed on either side of the tube and separated by a distance equal to their radius can provide a uniform magnetic field whose strength can be varied by changing the current, I, in the coils. B can be calculated from the radius of the coils, the number of turns and the current.

The electric field is switched on and this deflects the beam and the dot on the screen moves. The magnetic field is then adjusted until the dot returns to its original position.

When the beam is undeflected,

$$\text{Electric force} + \text{magnetic force} = 0$$

The size of the electric force $= Ee = \dfrac{V_p e}{d}$

The size of the magnetic force $= Bev = Be\sqrt{2e\,V_a/m}$

Therefore $Ee = Bev$ i.e. $E = Bv$

Since $E^2 = B^2v^2$ and $v^2 = \dfrac{2eV_a}{m}$

$$\therefore E^2 = \frac{2eV_aB^2}{m} = \frac{V_p^2}{d^2}$$

$$\Rightarrow \frac{e}{m} = \frac{V_p^2}{2B^2V_ad^2}$$

The voltages are measured with voltmeters, the distance d with vernier calipers or if already mounted in the tube with a ruler and a mirror to avoid parallax.

There are other methods and you should prepare the method with which you are familiar. One which is fairly simple to understand is the fine beam tube. Electrons are accelerated as above, but the space in which they move is filled with a gas at very low pressure. Their tracks can be seen in dim

lighting as a result of the gas molecules being excited. If a uniform magnetic field is applied they can be forced to move in circles, the radii of which can be measured using a ruler and a mirror to avoid parallax. The speed of the electrons is calculated as above.

The key equation in calculating the result is that the acceleration of the electrons, v^2/r, is equal to the magnetic force, Bev, divided by the mass, m.

(c) Speed from equation in (b)

$$= \sqrt{2eV_a/m} = \sqrt{2(1.76 \times 10^{11}\ \text{C kg}^{-1})(1.2 \times 10^3\ \text{V})}$$
$$= 2.1 \times 10^7\ \text{m s}^{-1}$$

There is no need to repeat the theory, but it is worth saying where you got the result from.

This speed is almost 10% of the speed of light and so the mass will be a little higher owing to relativistic effects. The higher mass will mean a slightly lower speed if the force is the same.

Example 7.5

Explain the following observations.

(a) A radioactive source is placed in front of a detector which can detect all forms of radioactive emission. It is found that the activity registered is noticeably reduced when a thin sheet of paper is placed between the source and the detector.

(b) A brass plate with a narrow vertical slit is now placed in front of the radioactive source and a horizontal magnetic field, normal to the line joining source and detector, is applied. It is found that the activity is further reduced.

(c) The magnetic field in (b) is removed and a sheet of aluminium is placed in front of the source. The activity recorded is similarly reduced.

(d) The aluminium sheet in (c) is replaced by a sheet of lead and the detector records much less activity. This activity is not affected by the reintroduction of the magnetic field. [7]

(AEB 1983)

Solution 7.5

(a) Thin paper absorbs alpha-particles, but allows beta-particles and gamma-ray photons to pass through. The radioactive source must be giving off an appreciable proportion of alpha-particles.

(b) The beta-particles are stopped by the brass plate, but can pass through the slit. The narrow beam of beta-particles is deflected by the magnetic field because they have a negative charge. The deflected beam misses the counter and thus the counter is now only recording the gamma-ray photons which are neither stopped by the brass nor deflected by magnetic fields.

(c) A couple of millimetres of aluminium sheet will stop all beta-particles. Thus the counter is still only recording gamma-radiation.

(d) The sheet of lead will stop all beta-radiation and stop some of the gamma-radiation, but not all of it. The gamma-ray photons that do pass through the lead are not deflected by the field, because they are uncharged.

Example 7.6

(a) (i) Discuss the statement that radioactive decay is a random process. [3]
(ii) Define the half-life $t_{1/2}$ and decay constant λ of a radioactive material. [4]
(iii) Deduce the relation between the two quantities. [5]

(b) How would you determine the half-life of a radioactive isotope for which the value is known to be of the order of minutes? Describe the experimental set-up, the procedure and how the result is calculated. [8]

What safety precautions would you take for an isotope known to emit β-radiation only? [2]

(c) The carbon-14 technique for dating archaeological materials depends on the assumption that when living organisms assimilate carbon ($^{12}_{6}C$) from the atmosphere they also assimilate some radioactive carbon-14 ($^{14}_{6}C$) atoms. The organism thus becomes slightly radioactive, emitting β-particles at a rate of 920 counts per hour for each gram of total carbon content in the organism. When the organism dies no more carbon is taken up and the carbon-14 present decays with a half-life of 5700 years.

In a dating experiment, 3.5 g of carbon from a sample of wood gave a count rate of 805 counts per hour, after allowing for background radiation. Estimate the age of the wood. [6]

(d) Background radiation normally has a mean value of 20 counts per minute that varies little from day to day. What are the sources of this natural background radiation? [2]

(OLE)

Solution 7.6

(a) (i) An unstable nucleus has a definite chance of decaying in a given time just as a die has a 1/6 chance of coming up five each time it is tossed. The die is random as is the nucleus.

If we had enough dice we should be able to say with some certainty that 1/6 of them would show five each time they were all tossed.

We are generally dealing with vast numbers of nuclei so we can be certain what will happen overall even though each event is a random one.

(ii) The half-life of a radioactive material is the time it takes for the unstable nuclei to halve in number.
The decay rate = λ × (the number of remaining unstable nuclei)
The activity of the source, $dN/dt = -\lambda N$

(iii) $N = N_0 e^{-\lambda t}$ and when $t = t_{1/2}$, $N/N_0 = 1/2$
$\therefore 1/2 = e^{-\lambda t_{1/2}} \Rightarrow e^{+\lambda t_{1/2}} = 2$
The relationship is $\lambda t_{1/2} = \ln 2$

(b) The radioactive material should be placed near a Geiger counter. The count rate should be recorded every ten seconds or so and the results plotted on a graph of count rate against time.

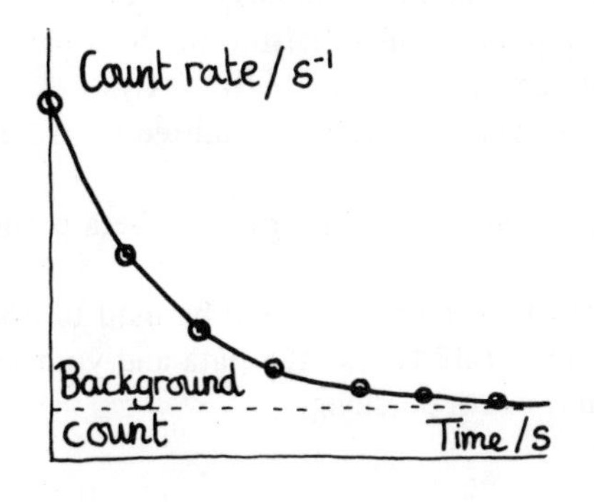

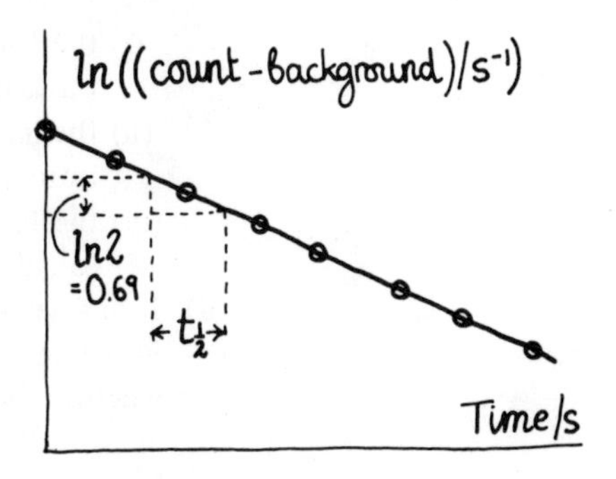

After many minutes the count rate will remain fairly steady. This is the background count.

Plot another graph of the natural logarithm of the amount by which the total count is higher than the background, against time. This graph will be a straight line and the half-life is the time it takes to drop by ln 2.

Plotting the straight line graph is a way of averaging out some of the random error.

A β-source should not be pointed at anyone, it should be handled with long forceps and kept in a lead box when not in use. Thick metal around the apparatus would protect the experimenter during the count.

(c) 3.5 g of carbon from new wood would give a count of

$$(3.5)\ (920 \text{ per hour}) = 3220 \text{ per hour}.$$

The 3.5 g of old wood gives a count of 805 per hour. This is 1/4 of the new wood count and so it is two half-lives old, that is 11 400 years.

Activity decays exponentially too because it is proportional to the number of unstable nuclei. It is possible to use a single formula to calculate the age of the sample:

$$\textbf{Age} = \frac{\textbf{ln (activity of new wood/activity of old wood)}}{\textbf{decay constant}}$$

$$= \frac{\textbf{half-life}}{\textbf{ln 2}} \times \textbf{ln (activity of new wood/activity of old wood)}$$

Most questions on decay use straightforward figures to avoid the necessity of candidates being very familiar with natural logarithms and exponentials.

(d) Background radiation comes from cosmic rays, natural radioactivity in the Earth and a tiny amount that is man-made.

Example 7.7

A radioactive isotope of thallium, $^{207}_{81}\text{Tl}$, emits beta particles ($\beta-$) with an average energy of 1.5 MeV. The half-life of the isotope is 135 days, and is thought to emit gamma radiation.

(a) (i) Describe simple tests which could be used to confirm that beta particles are emitted, and to check for the presence of gamma radiation. [5]

(ii) What will be the atomic number and the atomic mass of the new isotope formed by the emission of a beta particle? What will happen to the nucleus of the new isotope if a gamma ray photon is emitted? [4]

(b) (i) What is meant by an 'energy of 1.5 MeV', and what form does the energy take in this case? [2]

(ii) What is meant by a *half-life* of 135 days? [2]

(iii) Calculate the decay constant. [3]

(c) Assuming that 207 g of thallium 207 contains 6.0×10^{23} atoms, calculate

(i) the total energy, in joules, available from the beta particles emitted from 1 g of the isotope (electronic charge = -1.6×10^{-19} C);

(ii) the initial rate at which beta particles are emitted from 1 g of the freshly prepared isotope;

(iii) the initial power, in watts, available from the beta particles emitted at the rate calculated in (ii). [5]

(d) It has been suggested that thallium-207 could be used to power the amplifiers built into underwater telephone cables. Use the data and your answers to (c) to discuss whether the suggestion is worth pursuing. [4]

(AEB 1984)

Solution 7.7

(a) (i) The simplest check is that β-particles pass through paper which α-particles do not. But they are stopped by a few millimetres of aluminium which γ-rays are not.

If the count on a suitably placed GM tube and scaler does not fall to zero when paper is put in front of the source then β and γ may be present. If the count changes when aluminium is put in too, then β is present. If the count is still above the background count that would be there without the source, then γ is present.

(ii) A β-particle is ${}_{-1}^{0}e$ so

${}^{207}_{81}Tl = {}_{-1}^{0}e + {}^{207}_{82}X$, X has mass number 207 and atomic number 82

The emission of a γ-ray photon causes no change in the atomic or mass numbers, but the energy of the nucleus does decrease.

The mass number is also called the nucleon number and the atomic number is also called the proton number.

(b) (i) The eV is a unit of energy and 1 eV = 1.6×10^{-19} J.

The joule could be called a coulomb volt.

M means 10^6 so 1.5 MeV $= (1.5 \times 10^6)(1.6 \times 10^{-19}\text{ J})$
$= 2.4 \times 10^{-13}$ J

The energy is in the form of kinetic energy.

(ii) However many unstable nuclei there are, there will be half that number after a further 135 days.

(iii) $\lambda t_{\frac{1}{2}} = \ln 2$

$$\text{so } \lambda = \frac{0.693}{135 \times 86\,400\text{ s}} = 5.9 \times 10^{-8}\text{ s}^{-1}$$

(c) (i) The total number of β-particles available from 1 g of thallium
$= 6.0 \times 10^{23}/207 = 2.9 \times 10^{21}$

So the total energy available $= (2.9 \times 10^{21})(2.4 \times 10^{-13}\text{ J})$
$= 7.0 \times 10^{8}$ J

(ii) Initial rate of emission of β-particles
$= \lambda N_0 = (5.9 \times 10^{-8}\text{ s}^{-1})(2.9 \times 10^{21}) = 1.7 \times 10^{14}\text{ s}^{-1}$

(iii) Power $= (1.7 \times 10^{14}\text{ s}^{-1})(2.4 \times 10^{-13}\text{ J}) = 41$ W

(d) The power of 40 W is not yet electrical power. The conversion process is unlikely to be efficient.

Also the power source would be down to 1/8 of its already low power within 3 x 135 days which is about a year.

The idea is not worth pursuing.

Underlining the word *not* can avoid mistakes in writing and reading.

Example 7.8

A point source of alpha particles, a tiny mass of the nuclide ${}^{241}_{95}Am$, is mounted 7.0 cm in front of a GM tube whose mica window has a receiving area of 3.0 cm^2.

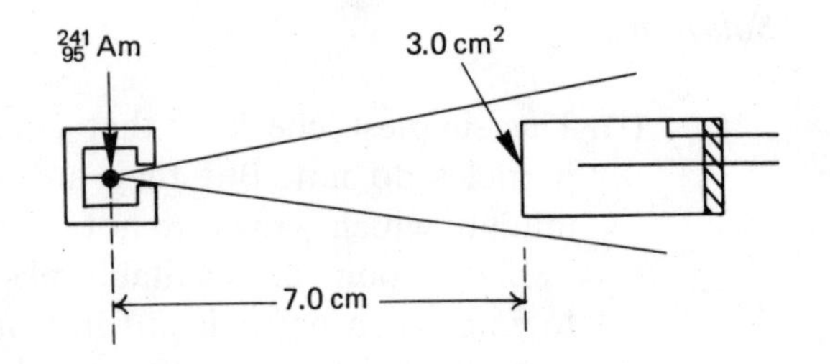

The counter linked to the GM tube records 5.4×10^4 counts per minute. Calculate

(a) the number of disintegrations per second within the source, and

(b) the number of $^{241}_{95}$Am atoms in the source.

The decay constant, λ, for $^{241}_{95}$Am $= 4.80 \times 10^{-11}\ s^{-1}$. [5]

(L)

Solution 7.8

(a) Assuming that each of the alphas that hit the mica window cause a single count,

$$\text{number of alphas hitting window per second} = 5.4 \times 10^4/60 = 900$$

The total area of a spherical surface 7.0 cm from the source $= 4\pi r^2 = 4\pi\,(7.0\ \text{cm})^2 = 616\ \text{cm}^2$. Only 3 cm^2 of this surface is counting alphas so the number of alphas being emitted in all directions $= (900\ s^{-1})\,(616/3.0) = 1.8 \times 10^5\ s^{-1}$.

It is important to be able to imagine things happening in three dimensions.

The number of disintegrations is therefore $= 1.8 \times 10^5\ s^{-1}$.

(b) Owing to the random nature of radioactive decay the disintegration rate, dN/dt, is proportional to the number of undecayed atoms, N, or $dN/dt = -\lambda N$. So

$$N = \frac{dN}{dt} \bigg/ \lambda = \frac{1.8 \times 10^5\ s^{-1}}{4.8 \times 10^{-11}\ s^{-1}} = 3.8 \times 10^{15}$$

Example 7.9

The diagram below represents components of the core of a nuclear reactor.

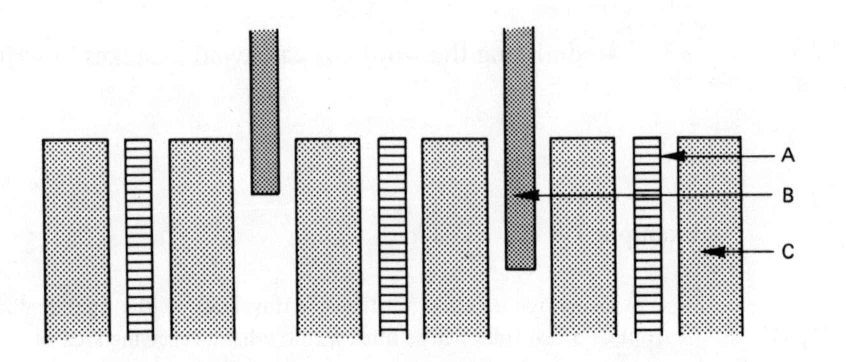

(a) Name the components represented by A, B and C in the diagram and explain their functions when nuclear energy is being continuously released in the reactor.
(b) Explain how this energy is removed from the core of an Advanced Gas-Cooled Reactor.
(c) Describe the further stages involved in using this energy to produce electricity. [11]
(SEB)

Solution 7.9

(a) A is a fuel element. It consists of enriched uranium in which $^{235}_{92}U$ nuclei react with slow or thermal neutrons – fission – to yield energy and further neutrons.
B is a control rod. It is made of a material which absorbs neutrons strongly, e.g. cadmium. By moving it in or out the number of fissions per second can be kept steady.
C is a moderator. It consists of material with a low proton number, e.g. graphite, which slows down the neutrons resulting from a fission and makes it possible for them to continue the chain reaction.

It would be easy to write a very long essay in answer to this part of the question, but you are only asked for names and functions. Parts (ii) and (iii) are best answered together.

(b) and (c) Carbon dioxide gas is pumped through pipes into the moderator. The hot CO_2 is then used to heat water, converting it to steam which in turn drives high and then low pressure steam turbines. These turbines are connected to the electrical generators – rotating electromagnets which induce e.m.f.'s in stationary coils.

Example 7.10

In the decay of radium-226 to radon-222 an α-particle of energy 4.95 MeV is released. Show that this is consistent with the principle of conservation of mass-energy given that the masses of $^{226}_{88}Ra$, $^{222}_{86}Rn$ and $^{4}_{2}He$ are respectively 226.0254 u, 222.0175 u and 4.0026 u, where $1\ u = 1.66 \times 10^{-27}$ kg and the speed of light = 3.0×10^8 m s^{-1}.

Solution 7.10

The mass loss in the reaction is equal to

$$226.0254\ u - (222.0175\ u + 4.0026\ u) = 0.0053\ u$$
$$= (0.0053)\ (1.66 \times 10^{-27}\ \text{kg}) = 8.8 \times 10^{-30}\ \text{kg}$$

***u* is the unified atomic mass constant sometimes referred to as an a.m.u.**

$$\text{Energy released} = mc^2$$
$$= (8.8 \times 10^{-30}\ \text{kg})\ (3.0 \times 10^8\ \text{m s}^{-1})^2$$
$$= 7.9 \times 10^{-13}\ \text{J}$$
$$= [(7.9 \times 10^{-13}) \div (1.6 \times 10^{-19})]\ \text{eV} = 5\ \text{MeV}$$

Note that you have to use 7 sig. fig. for the mass loss or mass defect calculation to get a result to 2 sig. fig. because the subtraction 'kills off' the first 5 figures.

Example 7.11

(a) (i) Describe the photoelectric effect. What are the principal features that are observed? [4]

(ii) State and explain Einstein's equation. How does it help interpret the mechanism of the photoelectric effect? [6]

(b) Outline the theory of an experiment, based on the photoelectric effect, for the determination of the Planck constant h. Practical details are not required. [8]

(c) An experimenter wishes to study the arrangement of planes of atoms in a crystal using the fact that a crystal acts as a diffraction grating for X-rays whose wavelength is of the order of the plane separation (10^{-10} m). Calculate the least accelerating voltage to be applied to an X-ray tube to produce suitable radiation. [6]

Take the Planck constant h to be 6.6×10^{-34} J s,
the speed of light *in vacuo* to be 3.0×10^{8} m s^{-1},
and the electronic charge e to be -1.6×10^{-19} C. (OLE)

Solution 7.11

(a) (i) Electromagnetic radiation falling on a clean metal surface can cause the emission of electrons from the metal.

If electrons are emitted the number per second increases as the brightness of the e.m. radiation is increased.

For a particular metal there is a frequency below which no electrons are emitted however bright the illumination.

As the frequency of the e.m. radiation is increased the kinetic energy of the emitted electrons increases and they can cross greater reverse potential differences.

(ii) $hf = \frac{1}{2}mv_{\text{max}}^2 + e\Phi$

All e.m. radiation is quantised into photons and each photon has an energy equal to its frequency f multiplied by the Planck constant h. m is the mass of the emitted electron, v its speed and e its charge. Φ is called the work function of the metal.

An energy equal to $e\Phi$ is required to release a single electron from a particular metal. If the energy of the photon interacting with the electron is less than this then nothing will happen. If $hf > e\Phi$ the surplus energy gives the electron more k.e. and thus it can cross a greater p.d.

One photon causes the emission of one electron and so if there are more sufficiently energetic photons there will be more electrons emitted and the current will be bigger.

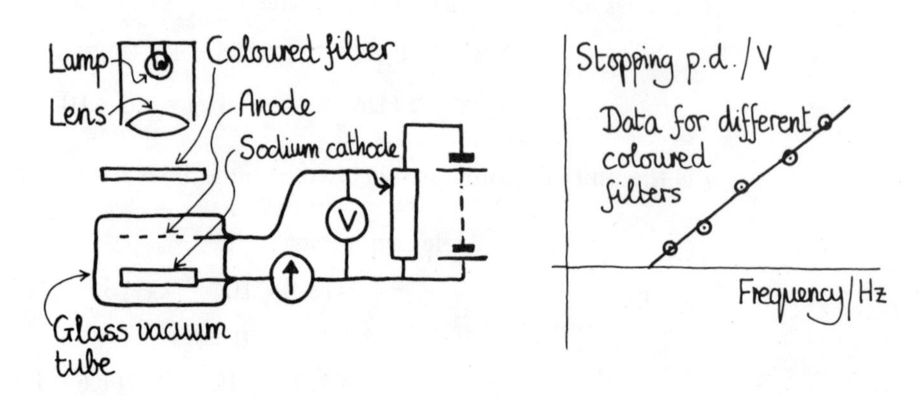

(b) The p.d. required to stop the emission of photoelectrons is plotted against the frequency of the light coming through the filter.

Since $\frac{1}{2}mv_{max}^2 = e \times$ (stopping p.d.)

the stopping p.d. $= hf/e - e\Phi/e = hf/e - \Phi$

The gradient of the graph is h/e and the y-intercept $-\Phi$.

h can be determined by measuring the gradient and multiplying by the charge on an electron.

(c) X-ray wavelength must be of the order 10^{-10} m

$$f = \frac{c}{\lambda}, \text{ so } f = \frac{3 \times 10^8 \text{ m s}^{-1}}{10^{-10} \text{ m}} = 3 \times 10^{18} \text{ Hz}$$

Energy of X-ray photon $= (7 \times 10^{-34} \text{ J s})(3 \times 10^{18} \text{ Hz})$
$= 2 \times 10^{-15}$ J

The electrons must have this energy if they are to cause the production of X-rays. $eV = 2 \times 10^{-15}$ J. Therefore

$$V = \frac{2 \times 10^{-15} \text{ J}}{1.6 \times 10^{-19} \text{ C}} = 1 \times 10^4 \text{ V, that is 10 kV}$$

Example 7.12

A very weak beam of yellow light, of wavelength 5.9×10^{-7} m, consists of photons which are separated, on average, by 0.20 m.

Explain why the average separation of the photons is given and not their actual separation and calculate the power carried by such a beam. [8]

Solution 7.12

Photons result from energy changes within atoms. These occur in a random way as atoms return to lower energy states after having been excited, e.g. by heating.

Each photon has energy $E = hf = \dfrac{hc}{\lambda}$

$$= \frac{(6.6 \times 10^{-34} \text{ J s})(3.0 \times 10^8 \text{ m s}^{-1})}{5.9 \times 10^{-7} \text{ m}}$$

$$= 3.4 \times 10^{-19} \text{ J}$$

As the photons travel at 3.0×10^8 m s^{-1}, then the number n passing a point in the beam every second is given by

$$n = \frac{3.0 \times 10^8 \text{ m s}^{-1}}{0.20 \text{ m}} = 1.5 \times 10^9 \text{ s}^{-1}$$

This calculation is a bit like $f = c/\lambda$ for waves – the units help to give confidence in the answer.

Therefore power $= nE$
$= (1.5 \times 10^9 \text{ s}^{-1})(3.4 \times 10^{-19} \text{ J})$
$= 5.0 \times 10^{-10}$ W or 0.50 nW

Example 7.13

(a) What do you understand by a line spectrum? [2]

(b) Explain why the existence of spectral lines supports the view that electrons in atoms exist in discrete energy levels. [5]

(c) The four lowest energy levels in a mercury atom are –10.4 eV, –5.5 eV, –3.7 eV and –1.6 eV.

(i) What is the ionisation energy of mercury? Give your answer in electronvolts and joules. [2]

(ii) Calculate the frequency and wavelength of the radiation emitted when an electron goes from the –1.6 eV energy level to the –5.5 eV level.

(iii) In what part of the electromagnetic spectrum does this wavelength lie? [6]

(iv) What is likely to happen if a mercury atom in the unexcited state is bombarded with electrons of energy

(A) 4.0 eV

(B) 6.7 eV

(C) 11.0 eV? [3]

Solution 7.13

(a) When the radiation from excited, isolated, atoms is viewed through a spectrometer that uses a prism or grating, lines are seen. These are images of the narrow slit with light of certain wavelengths. They indicate that the radiation is emitted in a number of very narrow bands of wavelength. The radiation from molecules or from solids and liquids is, by comparison, in wide bands or is continuous.

(b) The frequency of a photon is proportional to the size of the energy change that produces it. Hence each line in a spectrum corresponds to a certain energy change in the atom. Since there are only a certain number of lines there can only be so many changes possible. Therefore the electrons can only exist in certain energy levels. As an electron goes from one level to another lower level a photon of a certain energy is emitted.

(c) (i) The ionisation energy is the energy required to lift an electron from its lowest energy level to one corresponding to being completely removed from the atom. This highest level is by definition 0 eV.

Ionisation energy = 10.4 eV

$= (10.4 \text{ V}) (1.6 \times 10^{-19} \text{ C}) = 1.7 \times 10^{-18} \text{ J}$

(ii) Energy change $= (-5.5 \text{ eV}) - (-1.6 \text{ eV}) = -3.9 \text{ eV}$

$= (-3.9 \text{ V}) (1.6 \times 10^{-19} \text{ C}) = -6.2 \times 10^{-19} \text{ J}$

Energy of photon $= hf = 6.2 \times 10^{-19}$ J

$$f = \frac{6.2 \times 10^{-19} \text{ J}}{6.6 \times 10^{-34} \text{ J s}} = 9.5 \times 10^{14} \text{ Hz}$$

(iii) $$\lambda = \frac{c}{f} = \frac{3.0 \times 10^{8} \text{ m s}^{-1}}{9.4 \times 10^{14} \text{ Hz}} = 3.2 \times 10^{-7} \text{ m}$$

This is just below the wavelength of visible light, so the wavelength must lie in the ultra-violet region of the e.m. spectrum.

It is useful to have an idea of the figures in the e.m. spectrum for checking that answers are reasonable.

(iv) (A) 4.0 eV electrons do not have enough energy to lift the electron in the atom in the –10.4 eV level to the next level. Therefore nothing will happen to the atom.

(B) 6.7 eV electrons will excite the electron in the atom and lift it into the –3.7 eV state. (–10.4 + 6.7 = –3.7)

(C) 11 eV electrons will excite the electron in the atom so much that it will leave the atom. The atom will be ionised.

Example 7.14

A 150 W low pressure sodium vapour lamp produces a uniform energy flux of $0.20\ \text{W m}^{-2}$ of yellow light at a distance of 5.0 m from the lamp.

(a) Calculate

(i) the energy flux a distance 20 m from the lamp, and

(ii) the optical efficiency of the lamp, assuming only yellow light is emitted in the visible region of the electromagnetic spectrum.

(b) If each photon of yellow light from the lamp has an energy of about 3×10^{-19} J, estimate how many photons per second would strike the paper on which you are writing if it were held perpendicular to the light and 5 m from the lamp.

[7]
(L)

Solution 7.14

(a) (i) Energy flux obeys an inverse square law because the photons are being spread over larger spherical surfaces at greater distances from the lamp.

$$\text{Energy flux at 20 m} = \frac{(5.0\ \text{m})^2}{(20\ \text{m})^2}\ (0.20\ \text{W m}^{-2}) = 0.013\ \text{W m}^{-2}$$

(ii) The power is spread over a spherical surface of radius 5.0 m and hence area $4\pi\ (5.0\ \text{m})^2$.

Total power = $(0.20\ \text{W m}^{-2})\ 4\pi\ (5.0\ \text{m})^2 = 63$ W

The efficiency of the lamp = 63 W/150 W = 0.42 or 42%

Efficiencies are often in per cent, but do not have to be. Don't be muddled by the 100; it is best to look at the answer from a common sense point of view too.

(b) Area of a sheet of A4 paper = (0.2 m) (0.3 m) = $0.06\ \text{m}^2$

0.1 m^2 would be a good estimate.

Power arriving at paper = $(0.06\ \text{m}^2)\ (0.2\ \text{W m}^{-2}) = 0.012$ W

$$\text{Number of photons striking paper} = \frac{0.012\ \text{W}}{3 \times 10^{-19}\ \text{J}} = 4 \times 10^{16}\ \text{s}^{-1}$$

Do not put in more than 1 s.f. – it is an estimate!

7.8 Questions

Question 7.1

In the electric field in the electron gun of a cathode ray tube, an electron is speeded up from rest by a potential difference of 10 kV between the cathode and the anode.

Calculate the final speed of the electron. [3]

Question 7.2

The isotope $^{210}_{84}\text{Po}$ is found to decay with the emission of an α-particle to the stable isotope $^{206}_{82}\text{Pb}$ with a half-life of 140 days.

A scientist requires a laboratory source of $^{210}_{84}\text{Po}$ to have an activity between 2×10^4 Bq and 10×10^4 Bq (emissions per second). How often must he arrange for the source to be replaced? Show your reasoning. [4]

Question 7.3

A radioactive isotope of carbon, $^{14}_{6}C$, is produced in the upper atmosphere when $^{14}_{7}N$ absorbs a subatomic particle, X, and emits a proton. The Earth's upper atmosphere is rich in a variety of subatomic particles owing to its continual bombardment by cosmic radiation.

(a) Write a nuclear equation to represent the production of radioactive carbon from nitrogen-14 and hence deduce the nature of the absorbed subatomic particle X. [2]

The half-life of carbon-14 is about 6000 years. It is rapidly incorporated into the atmosphere's carbon dioxide and hence into the carbon cycle, so that any plant that absorbs carbon dioxide receives a proportional share of this radioactive carbon. When the plant dies the proportion of radioactive carbon to ordinary carbon goes down according to the decay law, and so the time that has elapsed since the death of the plant can be found.

(b) Sketch a graph to show how the ratio of radioactive carbon to ordinary carbon in a redwood tree which lived from 6500 B.C. to 6000 B.C. has varied with time from then until now. Put as much information as you can on the axes. [4]

(c) Suggest and explain one assumption on which this method of dating depends. [2]

Question 7.4

(a) Radium has an isotope $^{226}_{88}Ra$ of half-life approximately 1600 years. What is meant by the terms *isotope* and *half-life*? [2]

(b) A sample of $^{226}_{88}Ra$ emits both α-particles and γ-rays. State, and account for, any change in
   (i) nucleon number, A,
   (ii) proton number, Z,
   which may occur as a result of the emission of these radiations. [4]

(c) A mass defect of 8.8×10^{-30} kg occurs in the decay of a $^{226}_{88}Ra$ nucleus. Calculate the energy released.
   In a given sample it is found that most of the radium nuclei decay with the emission of an α-particle of energy 4.60 MeV and a γ-ray photon. What is the frequency of the γ-ray photon emitted? Ignore the recoil energy of the decayed nucleus. [5]

(d) Outline briefly how you could show experimentally that both α-particles and γ-rays are present in emissions from $^{226}_{88}Ra$. [5]
   How is it possible that, with a half-life of 1600 years, $^{226}_{88}Ra$ occurs in measurable quantities in minerals 10^9 years old? [2]
   (Speed of light = 3.0×10^8 m s^{-1}.
   The Planck constant = 6.6×10^{-34} J s.
   Electronic charge = -1.6×10^{-19} C.) (L)

Question 7.5

α-particles of energy 5.0 MeV from a radioactive source are used to collide with protons by firing the α-particles at a thin film of polythene in a vacuum chamber.

(a) In a head-on collision it was found that the speed of the ejected proton was 2.7×10^7 m s^{-1}. If the mass of a proton is 1.7×10^{-27} kg calculate
   (i) the change of momentum of the proton in the collision, stating any assumption made, and
   (ii) the initial momentum of the α-particle. You may assume that $m_\alpha = 4m_p$. [5]

(b) Use your answers to (a) to discuss the direction of motion of the colliding α-particle after the collision compared with before it. [2]

Question 7.6

Gamma rays are highly penetrating because of their weak interaction with matter; but when they do interact, with lead for example, the absorption process is complex. Nevertheless for a parallel beam of monoenergetic γ-rays a definite absorption law is followed.

The absorption of γ-rays by lead can be studied by placing an increasing thickness of lead between a pure source (e.g. 200 kBq cobalt-60) and a GM tube connected to a scaler, count rates measured and an absorption curve drawn. If the graph of ln(count rate) against absorber thickness is a straight line, the absorption law is exponential. Assuming this is so, we can say

$$C = C_0 e^{-ax}$$

where C_0 is the count rate produced by the γ-rays in the absence of absorbing material, C is the count rate with a thickness x of absorber interposed, and a is a constant called the linear absorption coefficient of the absorbing material.

(a) Sketch the arrangement of apparatus described for investigating the absorption of rays by lead.
What precautions would you take when performing the experiment? [4]

(b) (i) What does monoenergetic mean in the phrase 'a parallel beam of monoenergetic γ-rays'?
(ii) Suggest a typical value for the energy of a γ-ray photon. [3]

(c) Sketch the graph of ln C (up) against x (along) if the relationship between C and x is the form $C = C_0 e^{-ax}$.
Explain how, using your sketch, you would find the value of a, the linear absorption coefficient. [4]

(d) Gamma rays from a compact cobalt-60 source spread out in three dimensions. What would be the count rate on a GM tube held 0.60 m from the source if the reading at 0.20 m was 1200 s^{-1}? [2]

(e) It is found that a sheet of lead 4.0 mm thick reduces the count rate by a factor of two in the absorption experiment described, e.g. from C_0 to $C_0/2$ or from $C_0/3$ to $C_0/6$.
Use this information to calculate the linear absorption coefficient a. [4]

Question 7.7

(a) The first excitation potential for sodium gas is 2.11 V. Explain the term excitation potential and calculate the wavelength of the corresponding spectral line. [5]
Describe and explain what happens when a parallel beam of white light passes through a glass bulb filled with hot sodium vapour. [2]

(b) Describe briefly the phenomenon known as the photoelectric effect and explain the terms work function and threshold wavelength. [6]
The threshold wavelength for a clean sodium surface is 680 nm. Calculate
(i) the work function for sodium, and
(ii) the maximum kinetic energy for photoelectrons released by light of wavelength 390 nm.
(Speed of light in vacuum, $c = 3.00 \times 10^8$ m s^{-1}.
The electronic charge, $e = 1.60 \times 10^{-19}$ C.
The Planck constant, $h = 6.63 \times 10^{-34}$ J s.) [5]
(L)

7.9 Answers to Questions

7.1 Energy is being transformed from electrical to kinetic. The p.d. is the energy change per unit charge so one calculates the k.e. gained by an electron when it crosses a p.d. of 10 kV. The rest of the calculation is similar to that when calculating the speed of an object that has fallen a certain distance as a result of gravity.

Answer 5.9×10^7 m s^{-1}

This answer will not be quite right because at speeds approaching the speed of light the mass of the electron will be increasing. In this case it will increase by about 5%.

7.2 If he starts with a source of activity 10×10^4 Bq you can easily write down its activity after 140 days, 280 days, etc., and get a rough answer of about once a year.

The exact answer is 325 days.

7.3 This sort of question is essentially comprehension, that is understanding the question is half-way to answering it. You should (a) find that X is a neutron and (c) realise that the proportion of ^{14}C to ^{12}C in the atmosphere's CO_2 is assumed to be the same at all times. In (b) the graph will have both straight and curved parts. See worked example 7.6.

In fact the proportion does vary a little, but corrections can be made by counting the growth rings on living and dead trees such as the Bristle Cone Pine of North America.

7.4 (a) See a textbook.

(b) Accounting for a change in A or Z only involves say that, for example, Z goes down by 2 because an α-particle carries away 2 protons.

(c) You need to use $E = mc^2$ and, after converting 4.60 MeV to joules, find the energy of the γ-ray photon.

Answer 8.5×10^{19} Hz

(d) Absorption tests should be sufficient here; see worked example 7.5.

The last part depends on long-lived radioactive elements which decay into Ra-226. Look up radioactive series in your textbook.

7.5 (a) (i) 4.6×10^{-20} kg m s^{-1}, assuming the proton loses no energy moving through the polythene after the collision.

(ii) 10.4×10^{-20} kg m s^{-1}; see worked example 2.15. For the calculation going from energy to momentum first convert 5 MeV to joules.

(b) The α-particle does not change direction.

7.6 (a) Precautions for performing radioactivity experiments will be discussed in your textbook.

(b) (ii) Energy changes within the nucleus tend to be of the order of 1 MeV compared with 1 eV for energy changes in weakly bound orbital electrons.

(c) See the end of section 8.3 and worked example 7.6.

(d) 130 s^{-1}

(e) You need to use $\ln 2 = 0.69$

Answer 170 m^{-1} or 0.17 mm^{-1}

7.7 (a) 589 nm

Look up absorption spectra in your textbook to answer the second bit.

(b) A brief description would include reference to a simple demonstration experiment and a statement of what happens when you alter intensity *or* the wavelength of the incident light.

(i) 1.83 eV or 2.93×10^{-19} J

(ii) 2.18×10^{-19} J

8 Data Processing

8.1 You Should Appreciate

Throughout a physics course you will be continually required to analyse or process numerical data relating to the values of physical quantities. The data may have been obtained from your own experiments or projects or presented as the result of someone else's investigations. It may be given in the form of a table or a graph or even as a verbal list, and you should be able to translate information from one form to another.

Examination questions involving the processing of data will occur in all the different question types but will form a major component of many practical examinations and, for some Boards' examinations, form a section whose principal aim is to test this ability. In open-ended process questions the data may be incomplete or some may not be needed, at least for the answer you give. These questions are particularly hard and a perfect or complete answer is not possible, though full marks are; what you write becomes more a matter of opinion but must be supported by rough calculations and estimates.

Processing data is, of course, all about handling numbers but you must keep track of the units as physics is about physical quantities and not simply numbers: a p.d. of 0.63 is meaningless, it must be qualified by a unit and hence become, for example, 0.63 mV, i.e. 6.3×10^{-4} V or 0.63 kV, i.e. 630 V, etc. Chapter 1 gives more information on this point.

8.2 You Should be Able to Use

- A calculator having a full range of scientific functions:
  This will include finding arithmetic means, powers and roots, sines, cosines and tangents of angles in degrees or radians and their inverses, logarithms and exponentials. It is also an important skill to be able to do rough calculations to check calculator answers that seem unreasonable.
- Conventions for labelling graph axes and heading tables of data:
  These conventions are explained and examples given of their use on page 8. You will need to recognise and use the standard multiple and submultiple prefixes, e.g. M, mega or μ, micro. When plotting graphs decide whether or not the origin should appear and then choose a sensible scale. A good scale should make both plotting and reading easy.

8.3 Graphs

(a) Straight-Line Graphs

All straight-line graphs can be represented by the equation

$y = mx + c$

where m is the constant slope or gradient of the graph, $m = \Delta y/\Delta x$, and c is the intercept on the y-axis (the value of y when $x = 0$) (see Fig. 8.1). Only if $c = 0$ is y actually proportional to x. Both m and c will usually have units, e.g. if y is a p.d./V and x is a distance/m the slope m will have units of V/m (an electric field strength), and the intercept will have units of V (a p.d.).

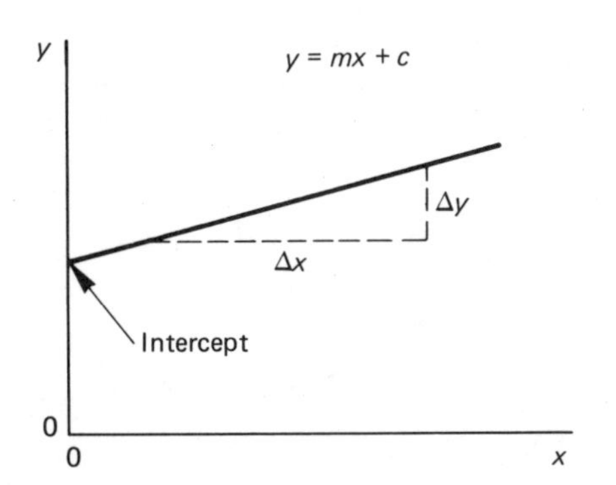

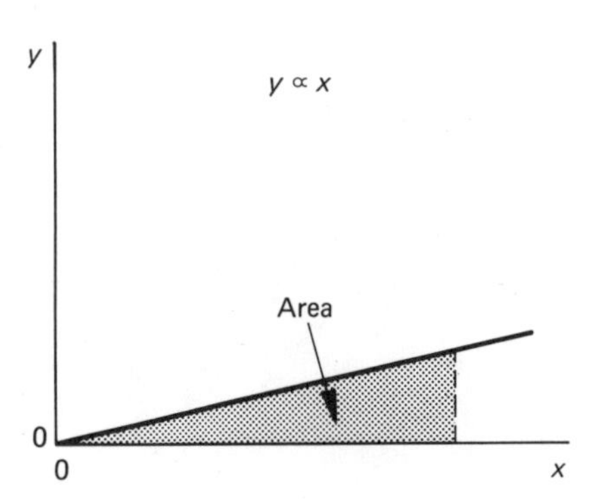

Figure 8.1

For curved graphs the slope at a point is the slope of a tangent drawn at that point. It can sometimes represent a useful physical quantity, e.g. the slope of a velocity–time graph represents the acceleration.

The area between a graph line and the x-axis sometimes represents a useful physical quantity, e.g. if y is a velocity/m s^{-1} and x is time/s, the shaded area (see Fig. 8.1) represents a displacement/m. The unit for the area is the units of y and x multiplied together.

When a non-linear relationship between two quantities is given and you are asked to plot data so as to produce a straight-line graph, it is necessary to recast the relationship in the form $y = mx + c$. For example $N = N_0 e^{-\lambda t}$ can be written as $\ln N = -\lambda t + \ln N_0$, so a graph of $\ln N$ as y against t as x will be a straight line of slope $-\lambda$ (see worked example 7.6).

(b) Curved Graphs

You should be familiar with several common graph shapes.

(i) $y = k/x$ *and* $y = k/x^2$

These inverse and inverse square relationships are met, for example, in Boyle's law and in gravitational field work respectively. Graphs like these do not touch either axis. See Fig. 8.2.

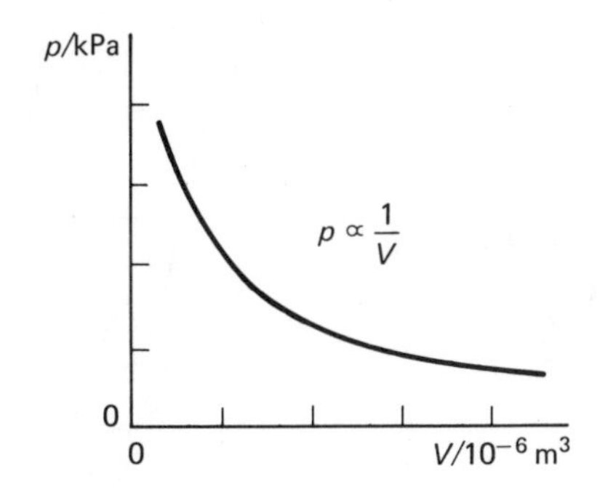

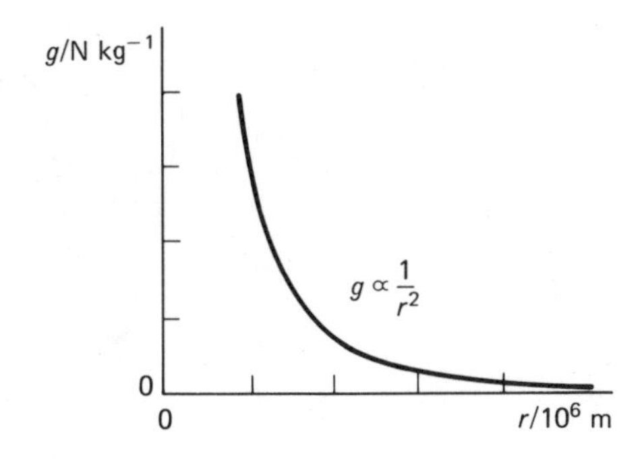

Figure 8.2

(ii) $y = ae^{-kx}$

This is exponential decay. Graphs like this start *on* the y-axis (unlike $y = k/x$) but do not touch the x-axis. Radioactive decay is a common example and as $\frac{dN}{dt} \propto N$ a graph of $\frac{dN}{dt}$ against t is also an exponential decay graph. See Fig. 8.3.

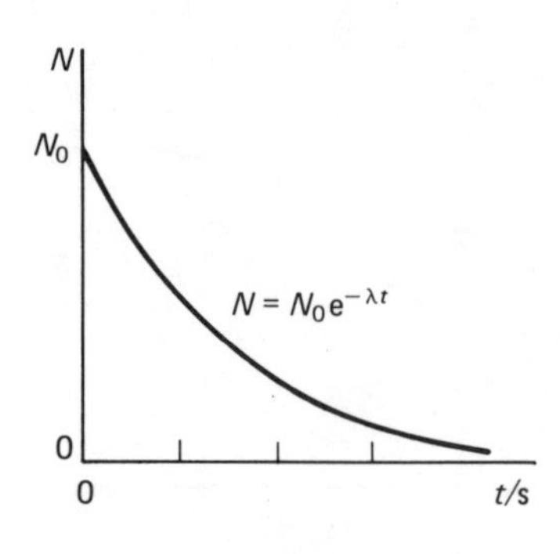

Figure 8.3

(iii) $y = kx^2$ *and* $y = ae^{kx}$

These represent a square relationship and exponential growth. The key difference is that the latter does *not* start at the origin but at $y = a$ when $x = 0$. The square relationship has initial slope of zero but the exponential does not. See Fig. 8.4.

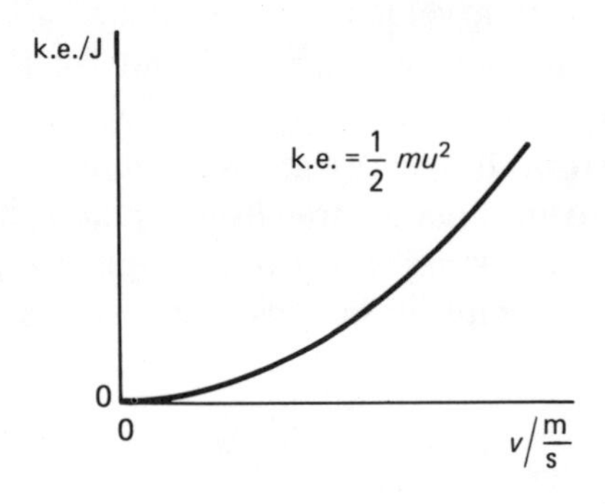

Figure 8.4

(iv) $y = y_0 \sin kx$ **and** $y = y_0 \cos kx$

These are trigonometric relationships. Both curves are sinusoidal, where they start being their only difference. The maximum slope for a curve of the form $y = y_0 \sin 2\pi ft$ or $y = y_0 \cos 2\pi ft$ is $2\pi f y_0$ and occurs as it crosses the time axis. See Fig. 8.5.

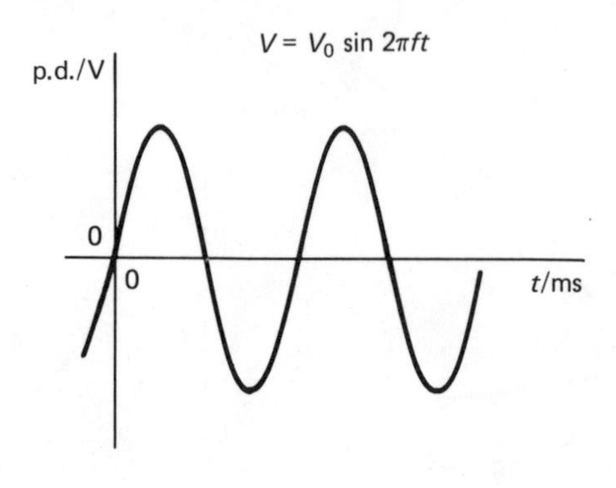

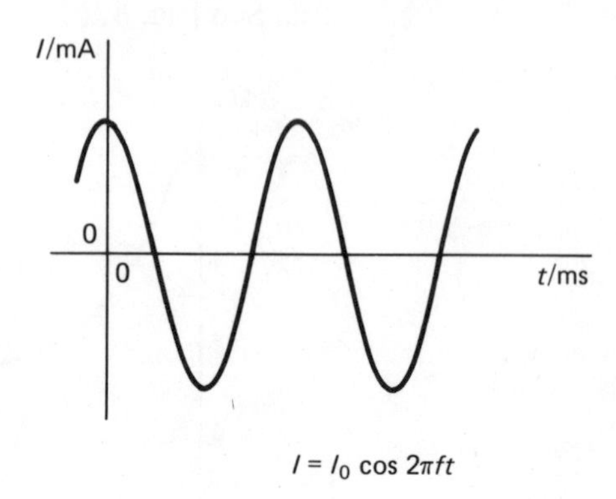

Figure 8.5

8.4 Errors

The uncertainty or *experimental* error in any measurement is an estimate and need only be given to one significant figure. To reduce uncertainties and avoid mistakes all measurements should be repeated.

The estimated uncertainty Δx in x is best expressed as a percentage, $(\Delta x/x) \times 100\%$. In using scales and thermometers or reading meters you can generally guess to about one tenth of a scale division with an uncertainty of two or three tenths of a scale division.

Measurements are also subject to *systematic* errors which result from an instrument being incorrectly calibrated or having a zero error. These errors are not revealed by repeating a measurement.

When several quantities are measured in an experiment the uncertainty in the final result will be dominated by the percentage uncertainty of the poorest measurement. In improving an experiment the poorest measurement should be repeated many times or the design of the experiment changed so as to obtain a series of results which can be processed graphically so as to average out random experimental errors.

8.5 Worked Examples

Example 8.1

A student, wishing to determine the area of cross-section of a wire, takes a single reading of its diameter with a micrometer screw gauge, obtaining a value of 0.20 mm. Assuming that his reading of the diameter is subject to an uncertainty of ±0.02 mm, which of the following is the most appropriate way of writing the result for the area, with its associated uncertainty?

A 0.03 ± 0.01 mm^2 **B** 0.031 ± 0.003 mm^2 **C** 0.031 ± 0.006 mm^2
D 0.0314 ± 0.0031 mm^2 **E** 0.0314 ± 0.0063 mm^2 (NISEC)

Solution 8.1

The area $\pi r^2 = \pi\,(0.01 \text{ mm})^2$
$= 0.0314 \text{ mm}^2$ to 3 sig. fig.

But the diameter of (0.20 ± 0.02) mm represents an uncertainty of 10% and this will also be the uncertainty in r.

As r is squared the uncertainty in r^2 is therefore 20%, and the uncertainty in πr^2 is ± 0.0063 mm^2.

The area can thus only be given to, at best, 2 sig. fig.

i.e. area = (0.031 ± 0.006) mm^2

Answer **C**

Example 8.2

A prototype car of mass 800 kg is being tested by running it at speed into an obstruction. The obstruction has measuring equipment built into it. The graph below shows how the push of the obstruction on the car varies with time.

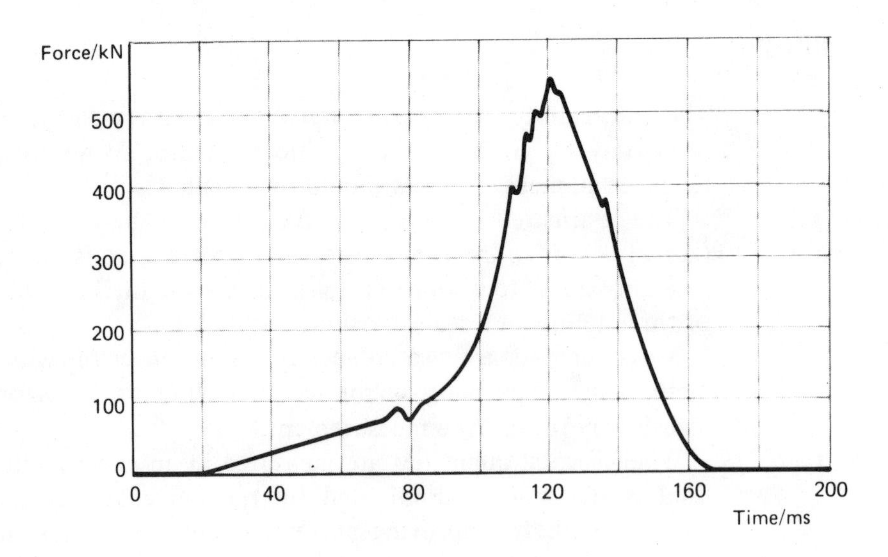

(a) Use the graph to estimate the change in momentum of the car on hitting the obstruction. [4]

(b) The wreckage of the car bounces off with a speed of 5 m s^{-1}. Calculate the speed of the car before the impact. [3]

Solution 8.2

(a) Each square under the graph represents an impulse of (100 000 N) (0.020 s) = 2000 N s.

There are about 13 squares so the total impulse or change in momentum = 13 x 2000 N s = 26 000 N s or 2.6 x 10^4 N s

Squares should be counted carefully but it would be foolish to claim too much accuracy in the method.

(b) The momentum after the collision = (800 kg) (–5 m s^{-1})
= –4000 kg m s^{-1}, or –4000 N s

The momentum before – (the momentum after) = 26 000 N s
i.e. the momentum before – (–4000 N s) = 26 000 N s
∴ the momentum before + 4000 N s = 26 000 N s
⇒ the momentum before = 22 000 N s

The speed before = 22 000 N s/800 kg = 28 m s^{-1}

The double negatives require great care. It is worth knowing that 28 metres per second is about 60 miles per hour.

Example 8.3

In an experiment to investigate the electrical properties of a filament lamp the following results were obtained for the p.d., V, across the lamp and the current, I, in it:

p.d./V	0.2	0.5	1.0	1.5	2.0	3.0	4.0	6.0	8.0	10.0	12.0
I/A	0.10	0.17	0.22	0.25	0.27	0.30	0.32	0.36	0.40	0.43	0.46

(a) Plot a graph of I (y-axis) against V (x-axis). Deduce from your graph the resistance of the lamp at room temperature. [6]

(b) Describe how the *resistance* of the lamp changes as the current in it increases. [5]

(c) (i) Add to your graph axes the I–V line for a wire of constant resistance 15.0 Ω.

(ii) If the lamp and resistor were joined in series, what p.d. would be needed to send a current of 0.2 A through them?

(iii) If the lamp and resistor were joined in parallel across a p.d. of 7.0 V, what current would be drawn from the source? [7]

Solution 8.3

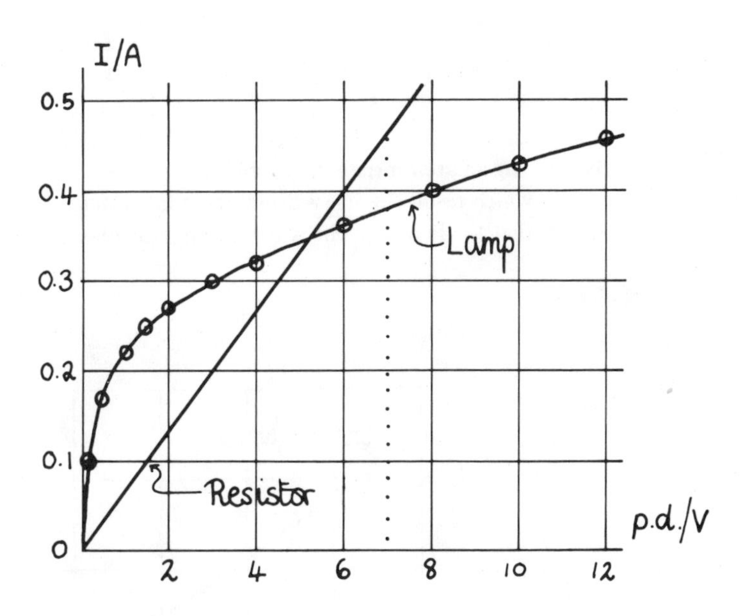

(a) The cold resistance = $\frac{V}{I}$ for very small currents

$$R_{\text{cold}} = \frac{0.2\ \text{V}}{0.10\ \text{A}} = 2\ \Omega$$

Though the value (0, 0) is not given in the question, the current in the lamp must be zero when there is no p.d. across it. The cold resistance can only be given to 1 sig. fig. as there is so little data at small currents.

(b) **The best approach here is to calculate some values and, perhaps, sketch a graph of *R* against *I*.**

I/A	0.1	0.2	0.25	0.3	0.35	0.4	0.45
R/Ω	2	4	6	10	15	20	25

The resistance is about 2 Ω for zero current and rises non-linearly as shown in the table.

(c) A 15 Ω resistor is described by a straight line from the origin through 3 V, 0.2 A or 6 V, 0.4 A, etc.

(i) From the graph a current of 0.2 A in the resistor needs 3.0 V and the same current in the lamp needs 0.7 V.
The total p.d. = 3.0 V + 0.7 V
= 3.7 V

(ii) From the graph a p.d. of 7.0 V would give a current of 0.47 A in the resistor and the same p.d. would give a current of 0.38 A in the lamp.
Total current = 0.47 A + 0.38 A
= 0.85 A

You should be able to quickly read off values for other cases.

More difficult questions based on graphs like this ask for estimates of the current when a p.d. of, for example, 6.0 V is connected across the lamp and resistor in series: it would be about 0.28 A.

Example 8.4

A student is investigating water flowing out of a burette. She records the level in the burette at various times and produces the table of results below.

time/s	0	5	10	15	20	25	30	40	50	60
height/cm	80.0	57.6	41.5	24.9	21.5	15.5	11.1	5.8	3.0	1.6

By plotting an appropriate graph show:

(a) which result she has written down incorrectly, [1]

(b) that the height of the water column decreases exponentially with time. [7]

Solution 8.4

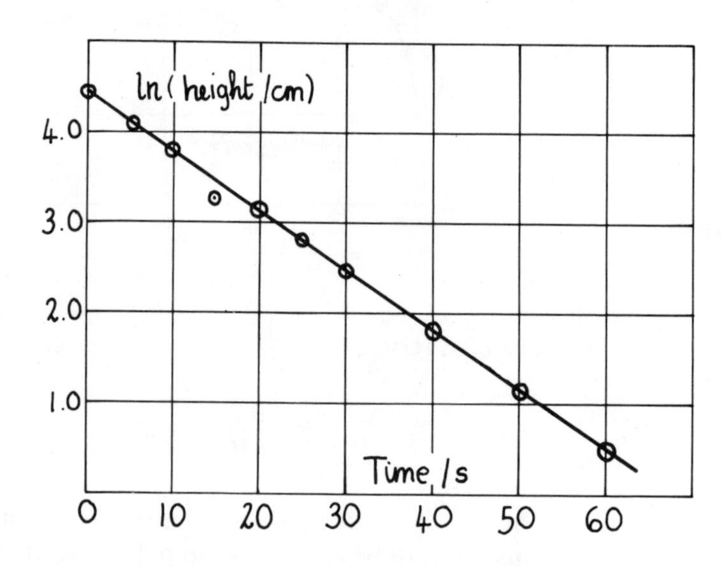

(a) The result at 15 s is quite out of line.

(b) If the height, $x = x_0 e^{-kt}$, where x_0 is the initial height, then $\ln x = \ln x_0 + (-kt)$.

Since $\ln x$ was plotted against t and a straight line results then x does decrease exponentially with time.

It would not be sufficient to plot x against t and show the graph looked exponential. A lot of functions might look roughly the same. The straight line clinches it.

Example 8.5

Particle accelerators can be used to accelerate charged particles to very high speeds. The kinetic energy, K, of a proton moving at a constant high speed v down a vacuum tube in the laboratory is given by

$$K = m_0 c^2 \left[\frac{1}{\left(1 - \frac{v^2}{c^2}\right)^{\frac{1}{2}}} - 1 \right]$$

where m_0 is the rest mass of the proton and c is the speed of light = 3.00×10^8 m s^{-1}. This expression for K reduces to

$$K \approx \tfrac{1}{2} m_0 v^2$$

for speeds which are small compared with c.

The table below gives the speeds, measured directly, for protons accelerated through given potential differences V over a wide range.

$V/10^8$ V	0.31	0.69	0.94	1.44	1.88	2.81	9.38	18.8	28.8	41.3
$v/10^8 \frac{\text{m}}{\text{s}}$	0.77	1.10	1.25	1.48	1.66	1.94	2.61	2.83	2.91	2.95

(a) Plot a graph of v^2 (y-axis) against K (x-axis) for values of v up to about 50% of the speed of light. Take the charge on a proton to be $+ 1.60 \times 10^{-19}$ C. [8]

(b) From this graph deduce a value for m_0. [4]

Solution 8.5

(a) A proton accelerated through 0.31×10^8 V acquires a kinetic energy given by

$$K = eV = (1.60 \times 10^{-19} \text{ C})(0.31 \times 10^8 \text{ V}) = 5.0 \times 10^{-12} \text{ J}$$

The question only asks for speeds up to about 1.5×10^8 m s^{-1} so a table of v^2 against K for the first five values of v will be enough.

$v^2/10^{16} \frac{\text{m}^2}{\text{s}^2}$	0.59	1.21	1.56	2.19	2.76
$K/10^{-12}$ J	5.0	11.0	15.0	23.0	30.1

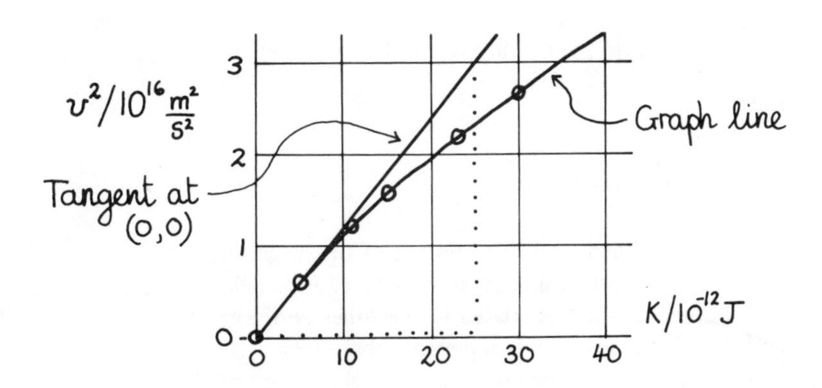

(b) As $K = \tfrac{1}{2} m_0 v^2$ then $v^2 = \dfrac{2}{m_0} K$

and the slope of the graph from the origin will be equal to $2/m_0$.

$$\text{Slope} = \frac{3.00 \times 10^{16} \text{ m}^2 \text{ s}^{-2}}{25.0 \times 10^{-12} \text{ J}} = 1.20 \times 10^{27} \text{ kg}^{-1}$$

$$m_0 = \frac{2}{1.20 \times 10^{27} \text{ kg}^{-1}} = 1.67 \times 10^{-27} \text{ kg}$$

The answer should perhaps be given to only 2 sig. fig. i.e. 1.7×10^{-27} kg. The units are correct as J = N m = kg m^2 s^{-2}.

8.6 Questions

Question 8.1

The following graphs show how one quantity, y, may vary with another quantity, x.

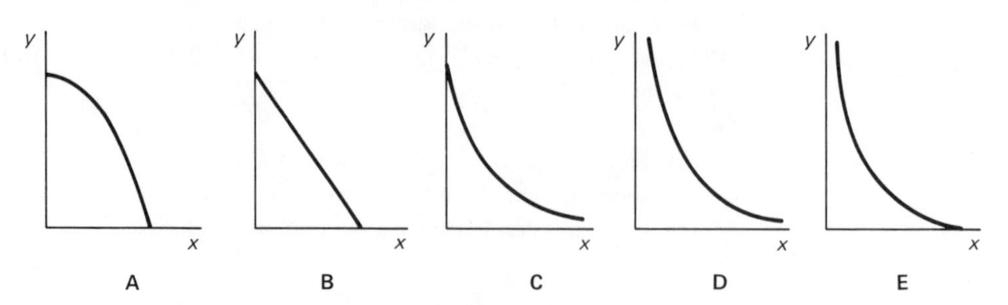

Which graph best illustrates the relationship between

	y-axis	x-axis
(a)	the velocity of a car during braking at constant acceleration	the time from the moment of applying the brakes
(b)	the mass of a rocket moving with a constant momentum	the velocity of the rocket
(c)	the frequency of an oscillating body	the time period of an oscillating body
(d)	the kinetic energy of a particle during a simple harmonic oscillation	the displacement of the particle from the centre of the oscillation
(e)	the reciprocal of the distance from a body to a camera lens	the reciprocal of the distance from the lens to the film
(f)	the current charging a capacitor through a resistor from a cell	the time from completing the circuit
(g)	the distance from an isolated charged sphere to a point	the electric potential at the point
(h)	the energy of a photon	the wavelength of the electromagnetic radiation associated with the photon
(i)	the number of undecayed nuclei in a sample of a radioactive element	the time from the preparation of the sample

Question 8.2

When a graph is plotted of the magnitude of a quantity y against the corresponding magnitudes of a related quantity x, areas below the graph give the magnitude of another quantity z which is related to the other two. The unit of z is also related to the units of x and y. For example, if y represents the velocity of a body moving in a straight line (unit: m s^{-1}) and x represents time (unit: s), the area z under the graph represents distance travelled (unit: m). In questions (a) to (d) below you are asked to identify the unit of the quantity represented by the shaded area below the curve, selecting the unit from the list (lettered **A** to **E**) below.

A joule **B** joule per kilogram **C** joule per cubic metre
D joule second **E** joule per second

For each graph described below, select the appropriate unit from the list above. Each response may be used once, more than once, or not at all.

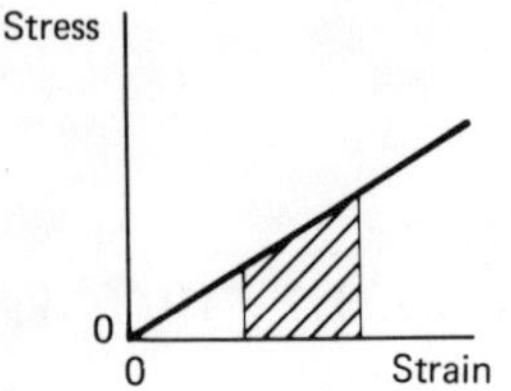

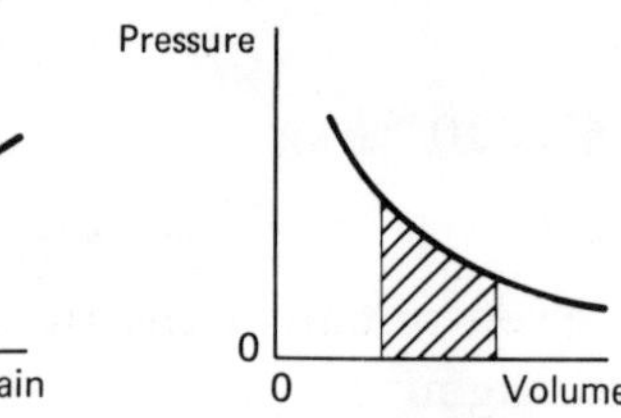

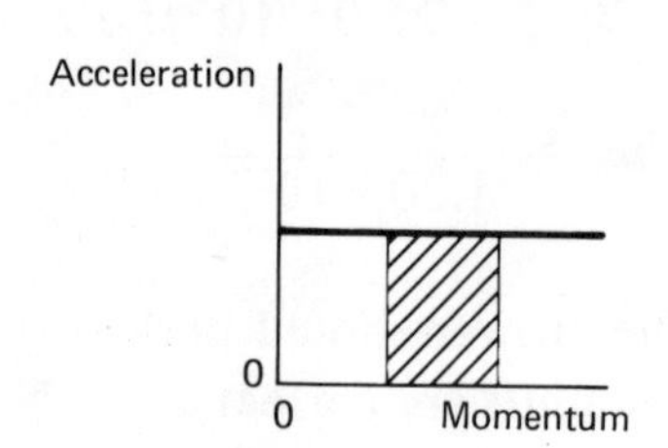

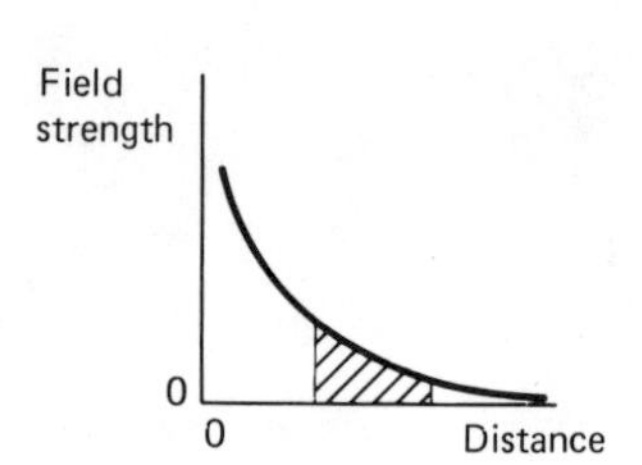

(a) The graph relating tensile stress in a rod and the corresponding strain.
(b) The graph relating the pressure and volume of a fixed mass of gas.
(c) The graph relating the acceleration of a body (moving in a straight line) and its momentum.
(d) The graph relating gravitational field strength and distance from a point mass giving rise to the field. (NISEC)

Question 8.3

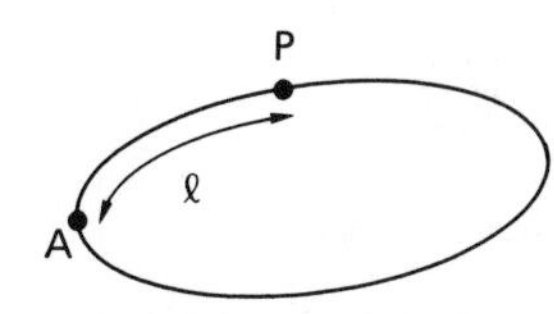

In an experiment the resistance, R, between A and P on a loop of wire of length L was measured for different positions of P. The length of wire, l, from A to P was also recorded and the results were:

l/m	0.20	0.40	0.60	0.80	0.90	1.10	1.20	1.40	1.60	1.80
R/Ω	11.6	21.1	27.6	31.6	32.6	32.5	31.5	27.6	21.0	11.9

The situation is essentially one of two resistors in parallel. If the wire has a resistance per unit length of k, then

$$\frac{1}{R} = \frac{1}{kl} + \frac{1}{k(L-l)}$$

$$\Rightarrow \quad L - l = \frac{RL}{kl}$$

(a) Plot a graph of l (y-axis) against R/l (x-axis) and from it deduce a value for (i) L and (ii) k. [10]
(b) Show how the second equation can be derived from the first equation and explain why the second equation is required. [4]

Question 8.4

Using the introduction and data given in the previous question
(a) plot a graph of R (y-axis) against l (x-axis). [4]
(b) The slope, S, of the graph in (a) is given by

$$S = k - \frac{2kl}{L}$$

Measure the slope at five values of l and, by plotting an appropriate graph, deduce values for k and L. [10]

Question 8.5

Choose two of the following crucial experiments and explain their significance by giving in note form
(a) the aim of the experiment;
(b) a labelled diagram of the apparatus;
(c) the observations made;
(d) the analysis and interpretation of the results.

A Rutherford's Scattering Experiment (done by Geiger and Marsden).
B Millikan's Oil Drop Experiment.
C Young's Two-slit Interference Experiment. [8]

Question 8.6

Using the introduction and data given in worked example 8.5

(a) plot a graph of K (y-axis) against $(1 - v^2/c^2)^{-\frac{1}{2}}$ (x-axis) for high values of v. Take the charge on the proton to be $+ 1.60 \times 10^{-19}$ C. [9]

(b) From this graph deduce two values for m_0

(i) from its slope, and

(ii) from the intercept on the K axis.

Comment on which value of m_0 is likely to have a smaller uncertainty. [5]

8.7 Answers to Questions

8.1 You should look carefully at how the graphs meet the y-axis and x-axis. Do they touch them or approach them without touching, for example? Try writing equations for each situation before deciding. **C** and **E** are the same shape with the axes interchanged.

Answers (a) **B**, (b) **D**, (c) **D**, (d) **A**, (e) **A**, (f) **C**, (g) **E**, (h) **D**, (i) **C**.

8.2 In the example given the area has units (m s^{-1}) (s) = m. For each of questions (a) to (d) you need to write down the units of y and x, multiply them together and then adjust them to include a joule: 1 J = 1 N m. The last question is the most difficult, especially if you start with (N kg^{-1}) (N s).

Answers **C, A, E, B**

8.3 (a) The graph is a straight line so it is not necessary to plot every one of the ten points, provided you use the ones at the extremes.

Answers (i) $L = 1.99$ m from the l intercept,

(ii) $k = 65.0\ \Omega\ \text{m}^{-1}$ from the slope.

(b) This is algebra pure and simple. The second equation involves two variable parts l and k/l which can be plotted to give a straight line.

8.4 (a) The graph is a parabola with its peak at $l = 1.00$ m.

(b) Drawing tangents and measuring slopes is time consuming. A graph of S (y-axis) against l (x-axis) will be a straight line. When $l = 0$, S is equal to k.

Answer $k = 65\ \Omega\ \text{m}^{-1}$, $L = 2.0$ m.

8.5 (a) Beware of overstating the aim. For instance

A is not to *prove* the existence of the nucleus,

B is not to measure the charge on an *electron*, and

C is not to prove that light *must* consist of waves,

though for two marks in an examination such statements may suffice.

In A and C, for example, the object is to test whether a certain model (of the nucleus or of light) might lead to a deeper understanding.

(c) and (d) Refer to your diagram and state how the observations lead back to the object in (a).

8.6 (a) The graph should be a straight line with a negative intercept on the K axis.

(b) You will have to 'see' the first equation for K in the form $y = mx + c$. Using the value for the speed of light given, the values for m_0 are

(i) from slope $m_0 = 1.67 \times 10^{-27}$ kg,

(ii) from intercept $m_0 = 1.7 \times 10^{-27}$ kg.

Answer (ii) is the more reliable.

Index